THE SOCIETY FOR APPLIED BACTERIOLOGY
TECHNICAL SERIES NO. 5

ISOLATION OF ANAEROBES

Edited by

D. A. SHAPTON

*H. J. Heinz Co. Ltd., Hayes Park, Hayes, Middlesex,
England*

AND

R. G. BOARD

*Department of Biological Sciences, University of Bath,
Claverton Down, Bath, England*

1971

ACADEMIC PRESS · LONDON · NEW YORK

ACADEMIC PRESS INC. (LONDON) LTD
BERKELEY SQUARE HOUSE
BERKELEY SQUARE
LONDON, WIX 6BA

U. S. Edition published by
ACADEMIC PRESS INC.
111 FIFTH AVENUE,
NEW YORK, NEW YORK 10003

Library of Congress Catalog Card Number : 70–149704
ISBN : 0–12–638840–7

Printed in Great Britain by
Cox & Wyman Ltd., Fakenham, Norfolk, England

Contributors

ELLA M. BARNES, *Food Research Institute, Colney Lane, Norwich, Norfolk, NOR 70F, England*

H. BEERENS, *Institute Pasteur, 20 Boulevard Louis XIV, Lille, France*

G. H. BOWDEN, *Dental Bacteriology and Biochemistry Department, The London Hospital Medical College, Turner Street, London, E.1, England*

G. S. COLEMAN, *Biochemistry Department, Agricultural Research Council Institute of Animal Physiology, Babraham, Cambridge, England*

J. S. CROWTHER, *Bacteriology Department, Wright–Fleming Institute, St. Mary's Hospital Medical School, London, W.2, England*

M. J. S. DADDS, *Process Research Department, Allied Breweries Ltd., Burton-upon-Trent, Staffordshire, England*

M. ELIZABETH DAVIES, *Royal (Dick) School of Veterinary Studies, Edinburgh, Scotland*

B. S. DRASAR, *Bacteriology Department, Wright–Fleming Institute, St. Mary's Hospital Medical School, London, W.2, England*

L. FIEVEZ, *Faculté de Médecine Vétérinaire, 45 Rue des Vétérinaires, Bruzelles Cureghem, Belgium*

S. M. FINEGOLD, *Wadsworth General Hospital, Veterans Administration, Los Angeles, California, 90073, U.S.A.*

B. W. F. FITZPATRICK, *Unilever Research Laboratory, Colworth House, Sharnbrook, Bedford, England*

BARBARA FREAME, *Unilever Research Laboratory, Colworth House, Sharnbrook, Bedford, England*

B. V. FUTTER, *Microbiology Section, School of Pharmacy, Portsmouth Polytechnic, Portsmouth, Hampshire, England*

A. C. GHOSH, *Food Hygiene Laboratory, Central Public Health Laboratory, Colindale Avenue, London, N.W.9, England*

J. M. HARDIE, *Dental Bacteriology and Biochemistry Department, The London Hospital Medical College, Turner Street, London, E.1, England*

E. HARRIS, *Department of Anaerobic Bacteriology, The Wellcome Research Laboratories, Langley Court, Beckenham, Kent, England*

BETTY C. HOBBS, *Food Hygiene Laboratory, Central Public Health Laboratory, Colindale Avenue, London, N.W.9. England*

G. HOBBS, *Torry Research Station, Department of Trade and Industry, Aberdeen, Scotland*

P. N. HOBSON, *Rowett Research Institute, Bucksburn, Aberdeen, Scotland*

C. S. IMPEY, *Food Research Institute, Colney Lane, Norwich, Norfolk, NOR 70F, England*

M. J. LATHAM, *National Institute for Research in Dairying, University of Reading, Reading, Berkshire, England*

S. O. MANN, *Rowett Research Institute, Bucksburn, Aberdeen, Scotland*

W. B. MOORE, *Department of Biochemistry, The Wellcome Research Laboratories, Langley Court, Beckenham, Kent, England*

J. A. MORRIS, *Microbiology Department, Reading University, Reading, RG1 5AQ, Berkshire, England*

EILEEN S. PANKHURST, *The Gas Council, London Research Station, Michael Road, London, S.W.6, England*

R. W. A. PARK, *Microbiology Department, Reading University, Reading, RG1 5AQ, Berkshire, England*

C. T. RICHARDSON, *Microbiology Section, School of Pharmacy, Portsmouth Polytechnic, Portsmouth, Hampshire, England*

M. ELIZABETH SHARPE, *National Institute for Research in Dairying, University of Reading, Reading, Berkshire, England*

F. A. SKINNER, *Soil Microbiology Department, Rothamsted Experimental Station, Harpenden, Hertfordshire, England*

P. T. SUGIHARA, *Wadsworth General Hospital, Veterans Administration, Los Angeles, California, 90073 U.S.A.*

VERA L. SUTTER, *Wadsworth General Hospital, Veterans Administration, Los Angeles, California, 90073, U.S.A.*

R. G. A. SUTTON, *Food Hygiene Laboratory, Central Public Health Laboratory, Colindale Avenue, London, N.W.9, England**

P. D. WALKER, *Department of Anaerobic Bacteriology, The Wellcome Research Laboratories, Langley Court, Beckenham, Kent, England*

R. WHITTENBURY, *Department of General Microbiology, University of Edinburgh, School of Agriculture, Westmains Road, Edinburgh, EH9 3JG, Scotland*

K. WILLIAMS, *Torry Research Station, Department of Trade and Industry, Aberdeen, Scotland†*

A. T. WILLIS, *Torry Research Station, Department of Trade and Industry, Aberdeen, Scotland†*

* Present address: Microbiology Department, School of Public Health and Tropical Medicine, University of Sydney, Sydney, Australia.

† Present address: Public Health Laboratory, Luton, Bedfordshire, England.

Preface

THIS volume includes contributions to the Autumn Demonstration Meeting of the Society for Applied Bacteriology, held on 22nd October, 1969 at the Department of Microbiology, Queen Elizabeth College, University of London. It is Number 5 in the Technical Series and it continues the Society's policy of providing experts in a particular field with the opportunity firstly of demonstrating methods and techniques to Members and guests of the Society and secondly of describing these in a book which is intended for use at the bench. Although this approach cannot be expected always to give complete coverage of a particular facet of microbiology, we feel that the efforts of the present contributors have produced a volume that describes a great number of techniques for the isolation of an extensive range of obligate or strictly anaerobic microorganisms from divers habitats. We wish to thank them all for the great effort which they took in the preparation of the Demonstration and their chapters in this book.

The Editors would like to express appreciation to Mr. A. Harry Walters for assistance in the preparation of this volume.

Our particular thanks go to Professor S. J. Pirt, Dr. G. Anagnostopoulos and other members of the staff of the Microbiology Department of Queen Elizabeth College for their help with the laboratory arrangements for the demonstrations.

December, 1970

D. A. SHAPTON
R. G. BOARD

Contents

Basic Methods for the Isolation of Clostridia

G. HOBBS, KATHLEEN WILLIAMS* AND A. T. WILLIS*

Torry Research Station, Department of Trade and Industry, Aberdeen, Scotland

Members of the genus *Clostridium* are anaerobic spore-forming bacteria. With a few exceptions none of them grow in the presence of air, although many species tolerate concentrations of oxygen below that present in the atmosphere.

No single method may be relied upon to isolate all the varieties of clostridia that might be present in a given sample, still less all the species that might be present in samples from widely different environments. A knowledge of the varying habits of this group of bacteria is, therefore, a prerequisite for their successful isolation. In the present communication, no attempt is made to describe all the specialized methods that have been devised to facilitate isolation of individual clostridia or groups of clostridia; rather, the general principles of isolation methods are described and illustrated by media and methods that we have found successful in examining specimens for the commoner mesophilic clostridia from a variety of environments including clinical material, fishery products, marine and fresh-water sediments. Details of other techniques are to be found in the publications of Smith (1955), Smith and Holdeman (1968), and Willis (1964, 1969).

Successful isolation of many bacteria depends on exploitation of specific properties of the organism sought. In the case of clostridia, anaerobic incubation in a liquid medium is itself often sufficient for effective enrichment, and isolation may be achieved simply by subsequent plating on to solid media. The relatively high resistance of spores to heat and chemicals may be exploited, provided the organism is present in the spore form; some clostridial spores, however, are no more resistant than many vegetative cells. Finally, various selective agents may be incorporated into media to inhibit the growth of other organisms. The success of individual selective agents depends, however, on the nature of the associated microflora, and no single inhibitor is universally effective.

* Present address: Public Health Laboratory, Luton

Anaerobic Conditions

Clostridia exhibit a wide range of tolerance to oxygen; some species will grow if only traces of oxygen remain, whereas a few species, notably *Cl. histolyticum* and *Cl. tertium* grow on agar media exposed to air. For the less exacting clostridia an adequately anaerobic atmosphere may be obtained in a sealed jar by evacuation and replacement of the air by either nitrogen or hydrogen. Indeed, use may be made of a partial anaerobic atmosphere for the isolation of relatively oxygen tolerant species such as *Cl. welchii* (Willis, 1964; Price and Shooter, 1964). Such conditions are obtained by removing the catalyst, evacuating the anaerobic jar to -60 cm of mercury and replacing the air with hydrogen. Such an atmosphere allows the growth of *Cl. welchii*, but not that of more strictly anaerobic, swarming species such as *Cl. sporogenes* and *Cl. tetani*.

Complete removal of oxygen from the atmosphere, which is required for the growth of the majority of clostridia, is achieved by the use of hydrogen and a catalyst, such as palladium, which converts any remaining oxygen in the jar to water. For the successful culture of very exacting species, such as *Cl. oedematiens*, even this procedure is not entirely satisfactory unless the culture medium itself is in a reduced condition (Moore, 1968). This is most conveniently achieved by using freshly prepared media containing 0·05% (w/v) cysteine hydrochloride.

Variations in the degree of reduction required by different species are illustrated in Figs 1–4. Growth of *Cl. tertium* and *Cl. histolyticum* occurs under aerobic conditions, but their growth is greatly improved by an anaerobic atmosphere. *Cl. tetani*, on the other hand, requires a strictly anaerobic atmosphere, whilst growth of *Cl. oedematiens* type D is unlikely to occur in an anaerobic jar unless the medium itself is reduced. We find it convenient to add 0·05% (w/v) cysteine hydrochloride routinely to all our plating media unless we are dealing with organisms of known relatively high oxygen tolerance.

Carbon dioxide often improves the growth of clostridia; it will certainly improve the germination rate of some spores (Wynne and Foster, 1948; Treadwell, Jann and Salle, 1958; Roberts and Hobbs, 1968). The carbon dioxide may be provided either in the atmosphere, for example as a mixture of hydrogen with 5% (v/v) carbon dioxide, or by the inclusion of 0·1% (w/v) of sodium bicarbonate in the culture medium.

It is pertinent to note that anaerobic growth of a Gram-positive spore-forming bacillus does not necessarily indicate the presence of a *Clostridium* spp. Many members of the genus *Bacillus* grow under anaerobic conditions producing a colonial morphology that closely resembles that of clostridia (Fig. 5). Indeed, almost any facultative anaerobe is likely to cause confusion

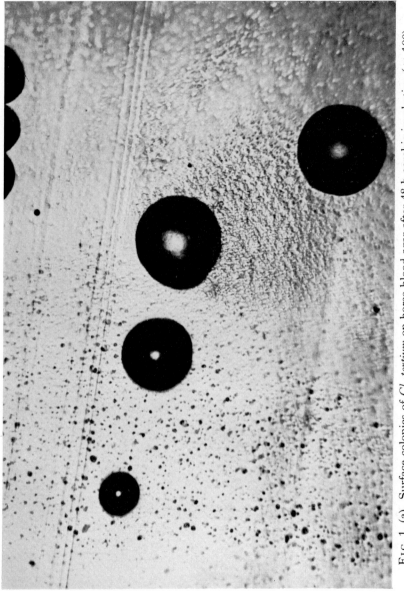

FIG. 1. (a). Surface colonies of *Cl. tertium* on horse blood agar after 48 h aerobic incubation (× 100).

B

Fig. 1. (b). Surface colonies of *Cl. tertium* on horse blood agar after 48 h anaerobic incubation ($\times$ 100).

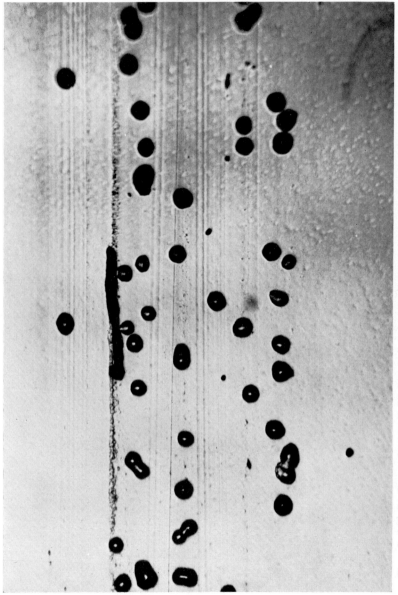

FIG. 2. (a). Surface colonies of *Cl. histolyticum* on horse blood agar after 48 h aerobic incubation (× 100).

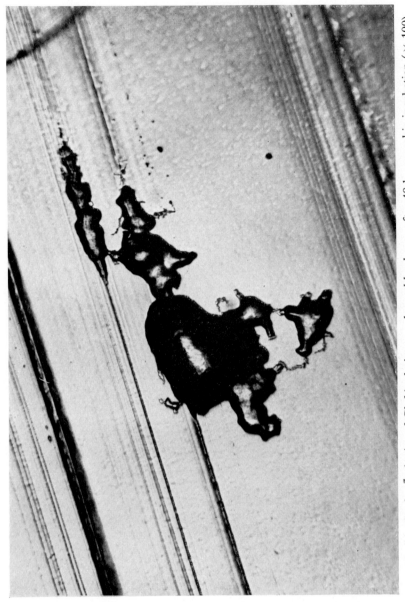

FIG. 2. (b). Surface colonies of *Cl. histolyticum* on horse blood agar after 48 h anaerobic incubation ($\times$ 100).

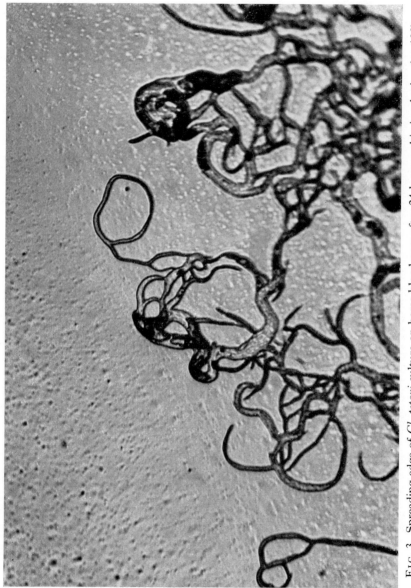

FIG. 3. Spreading edge of *Cl. tetani* culture on horse blood agar after 24 h anaerobic incubation ($\times$ 100).

in anaerobic work, and it is for this reason that every isolation method for any anaerobe must always incorporate a test for aerobic growth at every stage.

The final choice of method for achieving anaerobiosis is, to some extent, a matter of personal convenience. When a search is to be made for all

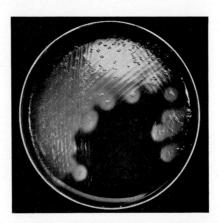

FIG. 4. Surface colonies of *Cl. oedematiens* type D on horse blood agar containing 0·05% cysteine hydrochloride. Culture incubated anaerobically for 48 h.

clostridia that might be present in a particular sample, incubation of cultures in an anaerobic jar equipped with a cold catalyst, and provided with a hydrogen/carbon dioxide mixture is the method we favour. On every occasion before use it is imperative that the anaerobic equipment should be tested to ensure that the jar does not leak, and that the catalyst is fully active—for details, see this volume p. 83. When this type of equipment is used, together with freshly prepared and adequately reduced media, little trouble is encountered in growing the most exacting species.

Treatment of Samples

Since the vegetative cells of many clostridia are sensitive to oxygen, it is important that samples should be exposed to air as little as possible. This implies rapid handling, and, if storage is necessary before processing, storage in the absence of air.

It is a common practice to treat samples with heat or chemicals before primary culture. The spores of many mesophilic clostridia survive heat treatment at 80° for 10 min, although survival is a function both of the number of spores present as well as of temperature and time of exposure. This treatment destroys all vegetative cells present. Various chemicals may

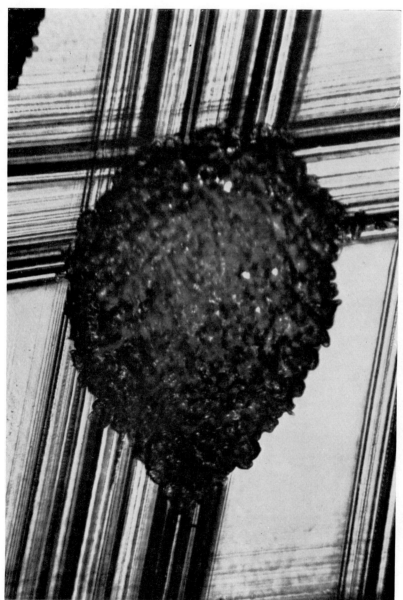

Fig. 5. (a). Surface growth of *B. cereus* on horse blood agar after 24 h aerobic incubation ($\times$ 200).

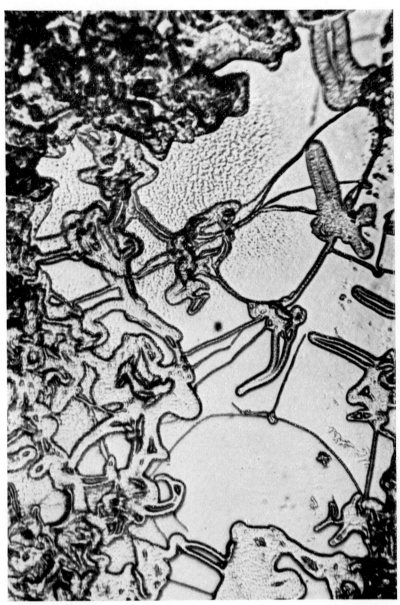

Fig. 5. (b). Surface growth of *B. cereus* on horse blood agar after 24 h anaerobic incubation (× 200).

also be used to kill vegetative organisms; indeed, use of chemicals such as ethyl alcohol may be successful when heating cannot be applied because the spores of some clostridia are thermolabile (Johnston, Harmon and Kautter, 1964). Thus, heat is appropriate for the isolation of *Cl. botulinum* type A, and ethanol treatment for the isolation of *C. botulinum* type E; the spores of the latter organism are sensitive to heat but not to alcohol treatment. Treatment of samples in the ways described assumes that the clostridia are present as spores; if they are present only as vegetative cells these methods cannot be used.

Culture Methods

Direct plating

Since clostridia are likely to be outnumbered by facultative anaerobes in many samples, direct plating of the sample on to solid media is of limited value. However, the ability of many clostridia to spread on the surface of moist agar media may often be exploited. Direct plating of samples for the isolation of *Cl. tetani* is a long-established and useful technique. Used alone, this method is of limited application, since the presence of other spreading organisms such as *Proteus* spp, renders it unsatisfactory unless the sample can be treated first, for example, by heating, to remove the non-sporing contaminants.

Enrichment culture

Primary enrichment in a liquid medium is commonly used for the isolation of clostridia. Although no single enrichment medium is universally acceptable, Robertson's cooked meat broth is suitable for the majority of clostridia, and is the one most generally used. This medium is efficient for both proteolytic and saccharolytic organisms, but it may be improved for the isolation of saccharolytic species by the addition to it of 0.5–1.0% (w/v) of glucose. When saccharolytic clostridia are sought to the exclusion of proteolytic species, enrichment in a maize mash medium may be used to advantage.

Selective agents

Addition of antibiotics or other selective agents to enrichment media may be found useful. Thus, the addition of neomycin sulphate (100 μg/ml) greatly favours growth of anaerobic organisms, and is especially useful when it is added to both the primary enrichment broth and the media used for subsequent plating (Lowbury and Lilly, 1955; Willis and Hobbs, 1959;

Willis, 1965). Phenethyl alcohol (0·25% v/v) may also be used with advantage for the inhibition of growth of Gram-negative facultative anaerobes (Dowell, Hill and Altemeier, 1964).

Incubation period

Because different clostridia do not grow at the same rate, the incubation time allowed before plating of the enrichment culture is important. For fast-growing species such as *Cl. welchii*, the enrichment culture should be plated out after 18–24 h incubation at 37° (or even earlier), whilst slower growing species such as *Cl. botulinum* and *Cl. tetani* are obtained by plating after 2–4 days' incubation, or longer.

Incubation temperature

The majority of clostridia grow at an optimum temperature of 37°, and unless thermophilic or psychrophilic species are being sought, this incubation temperature should be used. There are some occasions when temperatures other than the optimum may be useful. For example, *Cl. welchii* grows well at 45°, whereas many other mesophilic bacteria do not, and incubation of enrichment cultures at this temperature has been used successfully for the isolation of *Cl. welchii* from clinical material and water samples (Chapman, 1928; Wilson, 1938; Marshall, Steenbergen and McClung, 1965).

Plating of enrichment cultures

Because the isolation of clostridia is often aided by the recognition of specific reactions on solid media, the choice of media for plating of enrichment culture depends, to some extent, on the species being sought. The two most generally useful media are fresh blood agar and egg yolk agar. Most of the common clostridia may be isolated successfully on either of these, and in practice we have found it convenient to use both. A small number of species, however, such as the cellulose digesters and some of the saccharolytic pigmented species, do not grow well on these media.

Fresh blood agar

The choice of blood is important because haemolysins show varying activity against the erythrocytes of different animal species. This is well illustrated by the effect of different strains of *Cl. welchii* on horse blood agar and sheep blood agar. Only the θ-toxin of *Cl. welchii* is active against horse erythrocytes, whereas the α- and θ-toxins are both active against those of the sheep. Thus, whilst all strains of *Cl. welchii* type A produce α-toxin, by no means all of them produce θ-toxin, for example non-haemolytic

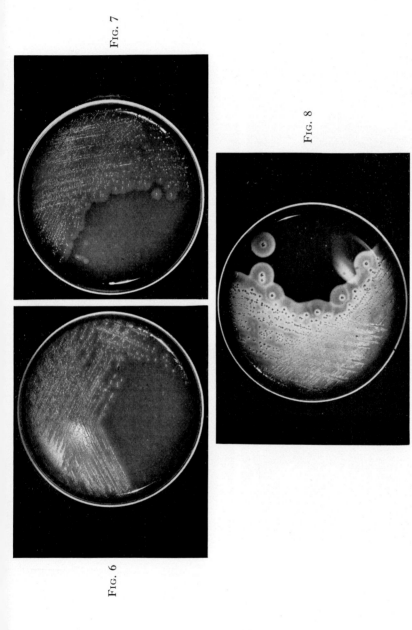

FIG. 7

FIG. 8

FIG. 6

FIGS 6–8. Cultures of a "non-haemolytic" food poisoning strain of *Cl. welchii* type A at 24 h on horse blood agar (Fig. 6), on horse blood agar containing 2% CaCl$_2$ (Fig. 7), and on sheep blood agar (Fig. 8). The organism produces little change on plain horse blood, but the α-toxin causes partial lysis of horse erythrocytes in the presence of CaCl$_2$. On sheep blood agar, there is extensive "hot-cold" haemolysis due to α-toxin activity.

food-poisoning strains; so that non-θ-producing strains are non-haemolytic on horse blood agar but are haemolytic on sheep blood agar. Extensive hot-cold haemolysis (α-toxin) is produced by strains of *Cl. welchii* growing on sheep blood agar. The addition of 0·5% (w/v) of CaCl₂ to horse blood

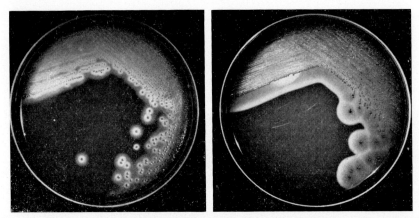

FIGS 9–10. Cultures of a classical type A strain of *Cl. welchii* at 24 h on horse blood agar containing 2% CaCl₂ (Fig. 9, left), and on sheep blood agar (Fig. 10, right). The appearance of "target" haemolysis on horse blood agar is due to the activities of θ-toxin (central complete lysis) and α-toxin (peripheral partial lysis). On sheep blood agar there is extensive haemolysis due to α- and θ-toxins.

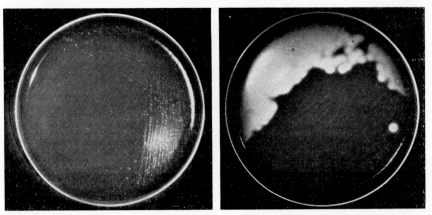

FIGS 11–12. Cultures of *Cl. chauvoei* at 24 h on horse blood agar (Fig. 11, left) and on sheep blood agar (Fig. 12, right).

agar renders the erythrocytes partially susceptible to α-toxin, so that colonies of classical type A strains produce "target" haemolysis on horse blood agar—a wide zone of partial haemolysis due to α-toxin and a narrow central zone of beta-haemolysis due to θ-toxin (Evans, 1945; Figs 6–10). Cultures of *Cl. chauvoei* produce only small zones of haemolysis on horse blood agar, but are strongly haemolytic on sheep blood agar (Figs 11–12).

Egg yolk agar media

There are several varieties of egg yolk agar in common use. For the isolation and recognition of many of the common clostridia, an egg yolk agar medium containing milk, lactose and a pH indicator is most generally useful (Willis and Hobbs, 1958, 1959). If the recognition of lactose fermentation is not required at this stage then the same medium, omitting the lactose and pH indicator, or alternatively the formulation of McClung and Toabe (1947), can be used. Apart from lactose fermentation and proteolysis, two reactions are readily recognized—lecithinase C and lipase activity. A lecithinase C reaction develops as a wide zone of opacity in the medium surrounding the colonies, and always extends well beyond their edge. The precipitate resulting from this reaction is composed chiefly of protein rendered insoluble by breakdown of lecithin in the medium (Willis and Gowland, 1962). A lipase reaction develops as a precipitate in the medium together with a co-extensive iridescent film (or pearly layer) that covers the colony. This reaction may extend a little beyond the edge of the colony, but is never as extensive as that due to lecithinases C. The lipase precipitate and the pearly layer are composed of free fatty acids formed by the breakdown of neutral lipids. The difference between the precipitates formed by lecithinases C and lipases may be readily demonstrated in a number of ways. Thus, the opacity resulting from lecithinase C activity is largely dissolved by trypsin, whilst the free fatty acids produced by lipolysis give an intense blue-green colour with saturated aqueous copper sulphate solution (Willis, 1960a, b; Willis and Gowland, 1962). These two types of reaction are exemplified by *Cl. welchii* (lecithinase C) and *Cl. sporogenes* (lipase) growing on lactose egg yolk agar (Figs 13–14). The reactions of some clostridia on lactose egg yolk agar and lactose egg yolk milk agar are illustrated in Figs 15–26.

As with most indicator media, these specific reactions on egg yolk agar may be adversely affected by minor variations in formulation of the medium. Thus, both the lecithinase and lipase reactions of clostridia may be modified by incorporation of an excess of utilizable carbohydrate. In the presence of much more than 1% (w/v) of fermentable carbohydrate the lecithinase activity of *Cl. welchii* is sometimes inhibited, an effect that is probably due to rapid fermentation of the sugar and consequent lowering of the pH. *Cl. botulinum* type E grown on egg yolk agar medium containing as little as 1% (w/v) of sucrose often fails to produce a pearly layer, although the precipitate in the medium is still present (Hobbs, Stiebrs and Eklund,

1967). Not infrequently this is then indistinguishable in appearance from a lecithinase C reaction. The mechanism of this inhibition is not understood, for both the pearly layer and the precipitate are composed of free

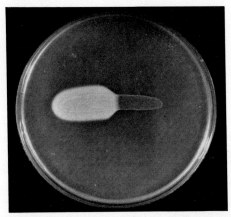

FIG. 13. Culture of *Cl. welchii* type A at 24 h on half-antitoxin egg yolk agar— *Cl. welchii* type A antitoxin on the right half of the plate. The diffuse lecithinase C opacity is inhibited by the α-antitoxin.

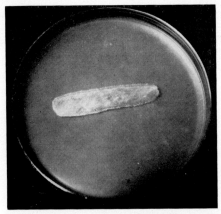

FIG. 14. Culture of *Cl. sporogenes* at 48 h on egg yolk agar showing the superficial "pearly layer" due to lipolysis.

fatty acids. This effect, moreover, cannot be produced with all lipolytic clostridia.

Both fresh blood agar and egg yolk agar may be rendered selective by the addition of antibiotics or other inhibitors. Neomycin sulphate, added to egg yolk agar or fresh blood agar at a concentration of 50–100 μg/ml, is

 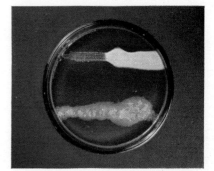

FIG. 15. (a) *Cl. welchii* type A.
(b) *Cl. bifermentans.*

FIG. 16. (a) *Cl. welchii* type C.
(b) *Cl. botulinum* type A.

 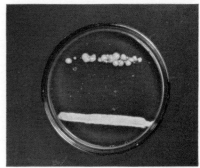

FIG. 17. (a) *Cl. œdematiens* type A.
(b) *Cl. sporogenes.*

FIG. 18. (a) *Cl. botulinum* type C.
(b) *Cl. sporogenes.*

 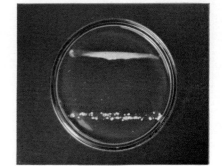

FIG. 19. (a) *Cl. histolyticum*
(b) *Cl. botulinum* type A.

FIG. 20. (a) *Cl. hæmolyticum.*
(b) *Cl. œdematiens* type B.

Streak cultures of clostridia on lactose-egg-yolk agar. In each case a mixture of
Cl. welchii type A and *Cl. œdematiens* type A antisera was spread on the left-hand
half of the plate.

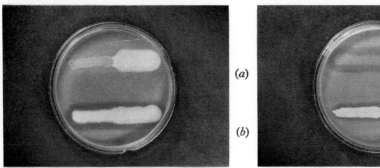

FIG. 21. (*a*) *Cl. welchii* type A.
(*b*) *Cl. bifermentans* (48 h).

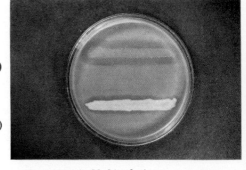

FIG. 22. (*a*) *Cl. histolyticum.*
(*b*) *Cl. sporogenes* (48 h).

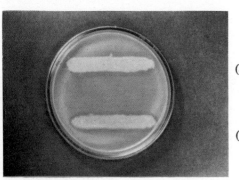

FIG. 23. (*a*) *Cl. botulinum* type A.
(*b*) *Cl. botulinum* type B (24 h).

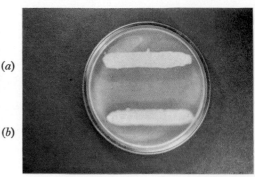

FIG. 24. (*a*) *Cl. botulinum* type A.
(*b*) *Cl. botulinum* type B (24 han-
aerobically and 48 h on bench).

FIG. 25. (*a*) *Cl. sporogenes.*
(*b*) *Cl. botulinum* type E (48 h).

FIG. 26. (*a*) *Cl. tertium.*
(*b*) *Cl. botulinum* type C (48 h).

Streak cultures of clostridia on lactose-egg-yolk-milk agar. In each case a mixture of
C. welchii type A and *Cl. œdematiens* type A antisera was spread on the left-hand half
of the plate. Fig. 4 shows the same culture as Fig. 3 after it had been on the bench
for 48 h.

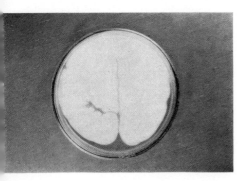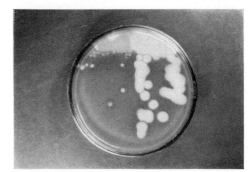

FIG. 27. Mixture of *Cl. welchii* type A and *B. cereus*.

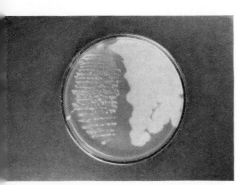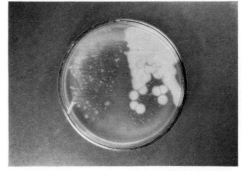

FIG. 28. Mixture of *Cl. welchii* type A, *B. cereus*, *Staph. aureus*, *E. coli*, *Ps. pyocyanea* and a *Proteus* species.

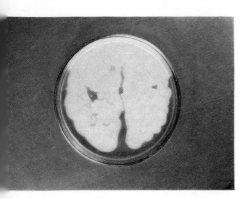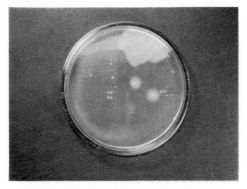

FIG. 29. Mixture of *Cl. welchii* type A, *Cl. sporogenes*, *Cl. histolyticum* and *B. cereus*.

Mixed cultures of aerobes and anaerobes on lactose-egg-yolk-milk agar (*left*) without and (*right*) with 250 µg/ml of neomycin sulphate. In each case a mixture of *Cl. welchii* type A and *Cl. œdematiens* type A antisera was spread on the left-hand half of the plate.

satisfactory for the routine isolation of most clostridia (Figs 27–29). Higher concentrations of the antibiotic may be used with advantage, especially when *quantitative* recovery of anaerobes is not required. Thus, isolation of *Cl. welchii* is readily made in the presence of 250 μg/ml of neomycin sulphate, but this concentration inhibits the growth of *Cl. botulinum* type E.

Firm agar media

The capacity of many clostridia to spread on the surface of agar media has been considered in relation to direct plating of samples. This property may also be exploited after growth of enrichment culture.

It is often profitable to inhibit spreading growth. Prevention of swarming not only enables isolated colonies of the spreading species to be obtained, but also facilitates isolation of non-spreaders whose colonies are otherwise overlaid by growth of the swarming organism. Spreading growth may be prevented in a number of ways, the simplest and most effective of which is the use of 3–4% (w/v) New Zealand agar in solid media instead of the usual 1·0–1·5% (w/v) agar (Miles, 1943; Hayward and Miles, 1943). The effect of using firm agar is well illustrated by comparing the growth of *Cl. tetani* on fresh horse blood agar containing 1·5 and 3·0% (w/v) of New Zealand agar respectively (Figs 30–31).

Another method that may be used to prevent the swarming growth of some organisms involves the incorporation of specific agglutinating antisera into agar media. This method has been used successfully in separating *Cl. tetani* and *Cl. septicum* from mixtures with other anaerobic organisms (Willis and Williams, 1970). The addition of 40–60 units/ml of commercial horse tetanus antitoxin (which contains significant amounts of agglutinating antibody) inhibits the swarming growth of *Cl. tetani*, and permits easy selection of colonies of other species present (Fig. 32).

Purification

Once growth of separate colonies of clostridia on one or another of the plating media has been achieved, it should not be assumed that these are composed of pure cultures. Colonies must be subcultured directly on to further agar media at least once more to exclude the possibility that they are composed of mixed species. In our experience, subcultures at this stage to fluid media, with subsequent re-plating, is much less successful in obtaining pure cultures. Even with direct plating, mixed cultures have been known to persist through many subcultures.

It is not uncommon to experience some difficulty in deciding which surface colonies should be selected for subculture. The colonies of many

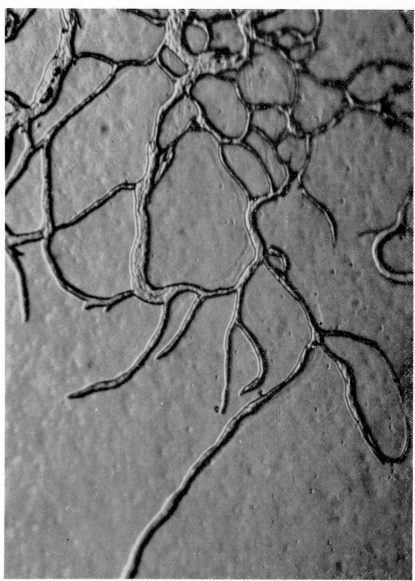

FIG. 30. The spreading rhizoidal edge of a culture of *Cl. tetani* on a 1·5% agar medium at 24h (× 200).

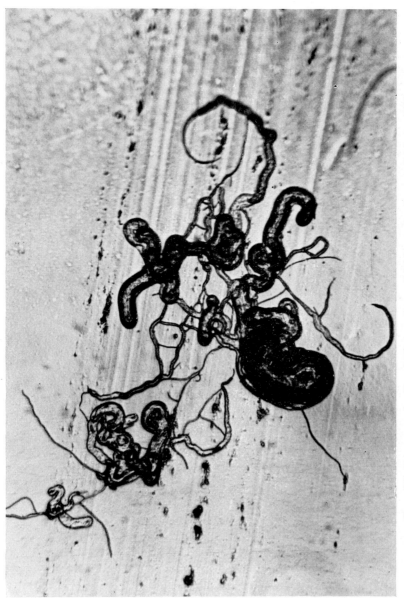

FIG. 31. Discrete colonies of *Cl. tetani* on a 3·0% agar medium at 24 h (× 200).

c

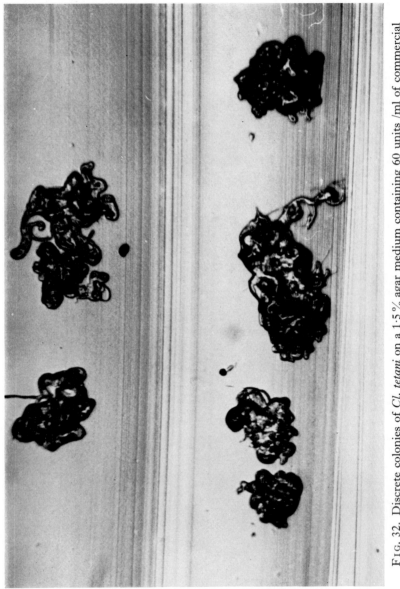

FIG. 32. Discrete colonies of *Cl. tetani* on a 1·5 % agar medium containing 60 units /ml of commercial tetanus antitoxin. Incubation period 24 h (× 200).

clostridial species closely resemble one another in young (24 h) cultures. In many cases, colonial differences are much more obvious after 2–4 days' incubation, but even at this time, there is a substantial number of species that produce a similar colonial morphology. Moreover, individual organisms in pure culture are likely to occur in more than one phase from time to time, especially in subcultures made from old fluid stock cultures. One of the commonest kinds of colonial variation of a pure culture is due to differences in the degree of sporulation in individual colonies; a colony with a high proportion of spores is much more opaque than one composed largely of vegetative cells.

Although colonial variation is not uncommon, the presence of colonies of different appearances should always be regarded **first** as indicating a mixed culture. Such colonies should be studied separately for microscopic morphology and cultural characteristics.

Appendix

Special reagents and methods

Cysteine hydrochloride

A fresh 10% (w/v) solution of cysteine hydrochloride (BDH) in distilled water is prepared. This is added to the agar base immediately before autoclaving, to give a final concentration of 0·05% (w/v). For exacting species of clostridia, media prepared in this way are used the same day.

Sodium bicarbonate

A 5% solution in distilled water is freshly prepared, sterilized by filtration and added to media to give a final concentration of 0·1% (2·0 ml to 100 ml medium).

Ethanol treatment

The sample containing clostridial spores is treated with 50% ethanol for 1 h at room temperature, and then treated in the usual way. If the sample is fluid, e.g. an enrichment culture, an equal volume of absolute ethanol is added to part of the sample. Non-fluid samples are emulsified in sterile distilled water to which an equal volume of absolute ethanol is then added.

Neomycin sulphate (Upjohn)

A solution of neomycin sulphate in distilled water is sterilized by filtration and added to sterilized media to give a final concentration of 50–100 μg/ml. Poured, uninoculated agar plates may be rendered selective by spreading a few drops of a 1% sterile aqueous solution of neomycin sulphate on the surface of the medium and allowing the plates to dry.

Media

Firm agar

Plate media are prepared in the usual way with New Zealand agar, except that the agar concentration used is 3 % (w/v). Because of the relatively high viscosity of the molten agar base, a little difficulty may be experienced in obtaining even distribution of additives such as egg yolk emulsion and blood.

Tetanus antitoxin agar

Ordinary fresh horse blood agar (1·0–1·5 % (w/v) agar) is prepared in the usual way, and to this is added equine tetanus antitoxin to give a final concentration of 40–60 units/ml. We have found it convenient to mix the antitoxin with the blood and to add this mixture to the cooled molten base.

Lactose egg yolk agar

1. The egg yolk is separated from the egg white by the usual culinary technique; no special aseptic precautions are necessary. An egg yolk suspension is prepared by mixing egg yolk with an equal volume (*c.* 20 ml/ yolk) of sterile 0·9 % saline solution.
2. A basal medium is prepared by mixing 100 ml of nutrient broth, 1·2 g New Zealand agar, 1·0 g lactose and 0·3 ml of a 1 % solution of neutral red. The mixture is autoclaved at 121° for 20 min. When this base medium has cooled to 50–55°, 4 ml of egg yolk suspension (1. above) are added. Plates are poured immediately.

Acknowledgements

The authors are indebted to Mr J. Harrison for his expert and careful production of the half-tone illustrations, and to the Editor and Publishers of the Journal of Pathology and Bacteriology for granting permission to use and for supplying the colour blocks. This chapter is the Ministry of Technology Paper for Publication No. T69/40 and was prepared as part of the programme of the Torry Research Station. Crown Copyright Reserved.

References

CHAPMAN, G. H. (1928). The isolation and estimation of *Clostridium welchii*. *J. Bact.*, **16**, 49.

DOWELL, V. R., HILL, E. O. & ALTEMEIER, W. A. (1964). Use of phenethyl alcohol in media for isolation of anaerobic bacteria. *J. Bact.*, **88**, 1811.

EVANS, D. G. (1945). The *in-vitro* production of α-toxin, β-haemolysin and hyaluronidase by strains of *Cl. welchii* type A, and the relationship of *in-vitro* properties to virulence for guinea-pigs. *J. Path. Bact.*, **57**, 75.

HAYWARD, N. J. & MILES, A. A. (1943). Inhibition of proteus in cultures from wounds. *Lancet*, **2**, 116.

HOBBS, G., STIEBRS, A. & EKLUND, M. W. (1967). Egg yolk reaction of *Clostridium botulinum* type E in different basal media. *J. Bact.*, **93**, 1192.

JOHNSTON, R., HARMON, S. & KAUTTER, D. (1964). Method to facilitate the isolation of *Clostridium botulinum* type E. *J. Bact.*, **88**, 1521.

LOWBURY, E. J. L. & LILLY, H. A. (1955). A selective plate medium for *Cl. welchii*. *J. Path. Bact.*, **70**, 105.

McCLUNG, L. S. & TOABE, R. (1947). The egg yolk reaction for the presumptive diagnosis of *Clostridium sporogenes* and certain species of the gangrene and botulinum groups. *J. Bact.*, **53**, 139.

MARSHALL, R. S., STEENBERGEN, J. F. & McCLUNG, L. S. (1965). Rapid technique for the enumeration of *Clostridium perfringens*. *Appl. Microbiol.*, **13**, 559.

MILES, A. A. (1943). Inhibition of swarming in plate culture. *Army Path. Lab. Serv., Curr. Notes*, No. **9**, 3.

MOORE, W. B. (1968). Solidified media suitable for the cultivation of *Clostridium novyi* type B. *J. gen. Microbiol.*, **53**, 514.

PRICE, D. J. E. & SHOOTER, R. A. (1964). Toxin production of faecal strains of *Clostridium welchii*. *Br. med. J.*, **2**, 1176.

ROBERTS, T. A. & Hobbs, G. (1968). Low temperature growth characteristics of clostridia. *J. appl. Bact.*, **31**, 75.

SMITH, L. DS. (1955). *Introduction to the pathogenic anaerobes.* Chicago: University of Chicago Press.

SMITH, L. DS. & HOLDEMAN, L. V. (1968). *The pathogenic anaerobic bacteria.* Springfield: Thomas.

TREADWELL, P. E., JANN, G. J. & SALLE, A. J. (1958). Studies on factors affecting the rapid germination of spores of *Cl. botulinum*. *J. Bact.*, **76**, 549.

WILLIS, A. T. (1960a). Observations on the Nagler reaction of some clostridia. *Nature, Lond.*, **185**, 943.

WILLIS, A. T. (1960b). The lipolytic activity of some clostridia. *J. Path. Bact.*, **80**, 379.

WILLIS, A. T. (1964). *Anaerobic bacteriology in clinical medicine.* 2nd ed., London: Butterworths.

WILLIS, A. T. (1965). Media for clostridia. *Lab. Pract.*, **14**, 690.

WILLIS, A. T. (1969). *Clostridia of wound infection.* London: Butterworths.

WILLIS, A. T. & GOWLAND, G. (1962). Some observations on the mechanism of the Nagler reaction. *J. Path. Bact.*, **83**, 219.

WILLIS, A. T. & HOBBS, G. (1958), A medium for the identification of clostridia producing opalescence in egg-yolk emulsions. *J. Path. Bact.*, **75**, 299.

WILLIS, A. T. & HOBBS, G. (1959). Some new media for the isolation and identification of clostridia. *J. Path. Bact.*, **77**, 511.

WILLIS, A. T. & WILLIAMS, K. (1970). Some cultural reactions of *clostridium tetani*. *J. Med. Microbiol.* **3**, 291.

WILSON, W. J. (1938). Isolation of *Bact. typhosum* by means of bismuth sulphite medium in water and milk-borne epidemics. *J. Hyg., Camb.*, **38**, 507.

WYNNE, E. S. & FOSTER, J. W. (1948). Physiological studies on spore germination with special reference to *Cl. botulinum*. III. Carbon dioxide and germination with a note on carbon dioxide and aerobic spores. *J. Bact.*, **55**, 331.

The Isolation of Clostridia from Animal Tissues

P. D. WALKER AND E. HARRIS

Department of Anaerobic Bacteriology

and

W. B. MOORE

Department of Biochemistry
The Wellcome Research Laboratories
Langley Court, Beckenham, Kent, England

The bacteriologist concerned with the isolation of clostridia from animal tissues is usually presented with material from a dead animal. Animals may have been dead for some time before discovery with the result that varying degrees of *post-mortem* change will have occurred.

On the death of an animal the natural barriers to infection are broken down and invasion of the tissues by organisms present in the intestine occurs. *Cl. welchii* type A and *Cl. septicum* are frequently recovered from the intestine of normal animals and as these organisms grow rapidly on the nutrient media commonly used for the isolation of anaerobes, they may rapidly outgrow the specific pathogen. The bacteriologist may thus be faced with isolating the causal organism from a mixture of contaminants. Techniques for the isolation of clostridia from animal tissues must take account of this factor. The following general principles should be observed.

(1) The tissues to be cultured must be carefully chosen.

(2) The medium used must be adequate for the growth of any pathogen.

(3) The tissue should not, in the first instance, be cultured in a fluid method but should be smeared directly on to the surface of the appropriate plates.

(4) Stiff agar or other anti-spreading devices should be used in order to minimize the spread of motile strains.

The different species of clostridia have different nutritional requirements and a variety of culture media are essential for satisfactory differentiation. Similarly, different species of clostridia differ in their sensitivity to oxygen and particular attention may have to be paid to the speed with which manipulations are undertaken and to the oxidation reduction potential of the culture medium. The growth of certain clostridia such as *Cl. oedematiens* types B, C and D is only achieved under strict anaerobic conditions. The reputation of the genus *Clostridium* for being a particularly difficult group to grow and isolate is due to neglect of the criteria outlined above.

In the livestock industry clostridial diseases are a major cause of economic loss. The organisms concerned are widespread and are found in the intestine of animals and in the soil, where they can both survive and multiply. Their ability to form spores ensures survival under adverse conditions for long periods. Pathogenicity is usually due to the production of exotoxins. Immunization programmes, either passive or active, are an integral and essential part of animal husbandry in many parts of the world. The principal clostridia causing disease in animals are:

Cl. welchii type B	Lamb dysentery; enterotoxaemia in calves, sheep, goats and foals.
Cl. welchii type C	Enterotoxaemia of sheep ("struck"), calves, lambs and piglets.
Cl. welchii type D	Enterotoxaemia in sheep, lambs and goats.
Cl. septicum	Gas gangrene; braxy in sheep; navel-ill in lambs; blackleg in pigs.
Cl. chauvoei	Blackleg in sheep, cattle and very occasionally pigs.
Cl. oedematiens type B	Black disease in sheep and "big-head" in rams.
Cl. oedematiens type D	Bacillary haemoglobinuria in cattle; necrotic liver disease in sheep.
Cl. sordelli	Gas gangrene from inoculation accidents.
Cl. tetani	Tetanus in all species of domestic animals.
Cl. botulinum types C and D	Botulism in sheep, cattle, mink and poultry.

Enterotoxaemias caused by organisms of *Cl. welchii* group usually follow changes of diet and husbandry. Sudden deaths in flocks associated with an increased level of feeding, such as occurs with the Spring flush of grass, is characteristic of *Cl. welchii* type D intoxication. In lambs up to three weeks of age, *Cl. welchii* type B can be responsible for widespread losses and has also been reported as a cause of death in single calves reared on their dams, thereby receiving excessive amounts of milk. In certain instances the disease may be restricted to a particular area, for example, "struck" in first-year ewes caused by *Cl. welchii* type C is associated with the Romney Marshes.

In the hill areas the occurrence of sudden deaths in flocks following the first heavy frosts is typical of braxy. This disease is thought to be precipitated by damage to the intestine by frozen food which results in the rapid multiplication of *Cl. septicum*. On *post-mortem* examination intense local inflammation in the wall of the fourth stomach (abomasom) is found.

Gas gangrene (Myonecrosis) in sheep and cattle is almost invariably caused by *Cl. chauvoei*, less commonly by *Cl. septicum* and rarely by *Cl. novyi* and *Cl. welchii* type A. Muscle damage due to bruising during handling or faulty inoculation of vaccine provides the conditions for spores circulating in the blood stream to localize and germinate. These spores presumably gain access to the circulation by passing across the intestinal wall or through the gums when teeth are erupting or being lost.

Similarly, damage to the liver by migrating immature flukes can provide suitable conditions for the germination of spores of *Cl. oedematiens* types B and D, the causes of black disease in sheep. Although the liver is regarded as the primary site of multiplication there is evidence that other organs may be involved (Batty, Buntain and Walker, 1964).

Cl. tetani is capable of causing disease in all domestic animals. Multiplication and toxin production are dependent on the presence of dead tissues resulting in anaerobic conditions and heavy losses have been experienced in flocks where lambs have been castrated or docked by the use of rubber rings. Horses are particularly susceptible and in some districts unvaccinated animals are frequently passively immunized before even minor surgical procedures are performed.

In the case of *Cl. welchii* infections the organisms can be readily cultured from the intestinal contents and from the liver and kidney. Intestinal contents should be preserved with chloroform (1 drop/10 ml) in those cases where there is a time lag before the material is received at the laboratory. Tissues for transport should be preserved in 50% glycerol saline. *Cl. septicum* organisms can usually be cultured from tissues taken from the abomasal wall in cases of braxy and in animals which have succumbed to blackleg, *Cl. chauvoei* can be isolated from pieces of the blackened muscle or rib marrow. *Cl. oedematiens* can be cultured from the liver, pericardial fluid or peritoneal fluid in cases of black disease and in cases of tetanus *Cl. tetani* can be cultured from the dead tissues associated with the rubber ring in animals castrated by this method.

It must be emphasized that the isolation of organisms from tissues should be interpreted with caution as a correct diagnosis can only be made in conjunction with a full clinical history. As already mentioned, large numbers of *Cl. welchii* and *Cl. septicum* can invade the tissues after death and as they grow readily on laboratory media can often be isolated from cases where they play no part in the pathogenesis. For this reason the fluorescent labelled antibody technique provides several advantages over the conventional cultural technique in that smears from the infected part show a distribution of organisms in the tissue at the time the smear was taken. Nevertheless, in badly decomposed tissues the results obtained with this technique are subject to the same limitations as the cultural method (Batty and Walker, 1963a, b).

Media

Blood agar will support the growth of most species of clostridia pathogenic for animals and is the routine medium of choice. All types of *Cl. welchii*, *Cl. sordellii*, *Cl. septicum* and *Cl. oedematiens* type A will grow readily on this medium. Only *Cl. chauvoei*, *Cl. botulinum* (types C and D) and *Cl. oedematiens* (types B and D) require special media.

The media used in these laboratories has been described in a separate communication (Batty and Walker, 1965) and is shown in Table 1. Media which are invariably based on a meat extract with added peptone, should be

TABLE 1. Media* for the isolation of clostridia from animal tissues

Normal blood agar:	nutrient broth containing 1·8% of agar (New Zealand agar ≡ 1·6 times Oxoid no. 3 is used throughout), 75 parts; pancreatic autodigest, 25 parts; horse blood, 5 parts, by volume.
Stiff blood agar:	as above but containing 3% of agar.
Cl. chauvoei medium:	nutrient broth containing 1·8% of agar, 75 parts; 50% glucose, 2 parts; liver extract, 3 parts; sheep's blood, 5 parts, by volume.

* Additional details given in text.

used as soon as possible after preparation. The peptone may be prepared quite simply as a papain digest of meat by the following method:

One litre of tap water is added to 1 kg of fresh minced muscle and stirred. Any fat which rises to the surface is skimmed off and 3·5 g of papain powder is added to the mixture. The temperature is raised to 60° for 2 h after which the enzyme activity is destroyed by boiling for 20 min. The mixture is then filtered through paper until clear. Total Nitrogen is measured by the Kjeldahl procedure and the appropriate amount is added to the meat extract; it is our experience that a 3% solution of "commercial peptone" has a Total Nitrogen content of approximately 5 g/litre. After addition of the peptone the pH is adjusted to 7·8 and the medium boiled to bring down phosphates. After removal of phosphates by filtration, agar can be added to give the complete medium.

In the case of *Cl. chauvoei* the nutrient agar is supplemented with liver extract and glucose (Batty and Walker, 1965). The liver extract is prepared as follows.

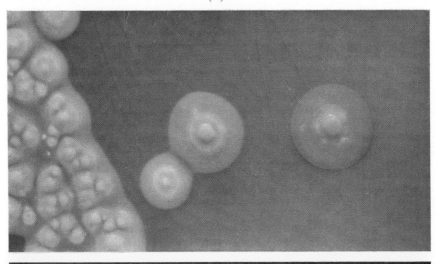

(A)

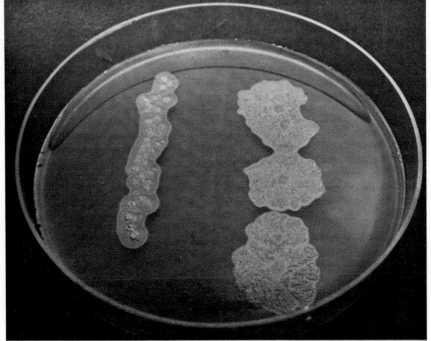

(B)

Fig. 1. (A) Colonies of *Cl. welchii* grown on lecithinase agar plates showing zones of opacity surrounding the colonies.
(B) Colonies of *Cl. botulinum* type E showing the pearly layer.

A sheep's liver is finely minced, covered with tap water and boiled for 30 min. After straining off the coarse particles through muslin, the supernatant is Seitz filtered and stored at 4°. It is advisable to make the extract fresh each week.

The medium used for *Cl. oedematiens* types B and D is described in the section dealing with the isolation of these organisms.

If it is desired to combine isolation with a rapid means of identification, lecithinase agar plates are very useful. In this medium the blood is replaced by an egg yolk solution made in the following way.

The yolk from one egg is carefully separated and emulsified in 25 ml of sterile saline (0·85% NaCl), the large particles allowed to settle and the supernatant withdrawn. The supernatant is used at a concentration of 5% (v/v) in nutrient agar and plates prepared. The egg yolk solution should not be kept for longer than 48 h, otherwise contaminants from the egg may become a nuisance. Strains of *Cl. welchii, Cl. bifermentans, Cl. sordellii* and *Cl. oedematiens* types A, B and D produce a zone of opalescence on this medium due to lecithinase activity (Fig. 1A). The lecithinases of *Cl. welchii* and *Cl. bifermentans/sordellii* are antigenically related and thus inhibited by antiserum to *Cl. welchii* type A. The lecithinases of *Cl. oedematiens* on the other hand are specific and inhibited only by their own antisera. The lecithinase activity of *Cl. oedematiens* type A is due to γ toxin and that of types B and D to β toxin. Strains of *Cl. sporogenes, Cl. botulinum* and *Cl. oedematiens* type A (ϵ toxin) produce a zone of opacity due to lipase activity and the colony is covered by a pearly layer (Fig. 1B). This effect is not inhibited by antiserum.

The inhibition of spreading

As pointed out above, increasing the agar concentration is the only method of inhibiting spreading which does not significantly inhibit growth of the organism as well. *Cl. tetani, Cl. septicum* and *Cl. sordellii* are the organisms most prone to spread on 1·8% blood agar. In any cultures in which the presence of these organisms is suspected it is necessary to use a 3% agar concentration if separate colonies, vital for purification, are to be obtained. The appearance of other species on stiff agar is somewhat altered and, therefore, increased agar concentration is not to be recommended for routine use as it will be difficult to correlate the appearance of colonies with those described in the literature.

Anaerobiosis

Although many methods have been described for obtaining anaerobic conditions (see Willis, 1960), these are of historical interest only and in practice the anaerobic jar is the method of choice. The present anaerobic jar (Baird and Tatlock Ltd.) is compact, convenient and reliable (for additional details, this volume p. 81). The frequency with which the catalyst has to be changed will depend on the amount of use and the nature of the organisms grown in the jar. It has been standard practice in our laboratory to place a second container of catalyst in the bottom of the jar and to change these in rotation at three-monthly intervals. A methylene blue indicator in agar is placed in a tube on the side arm of the jar, and this must also be changed frequently since the dye soon irreversibly bleaches and the agar dries at the surface thus preventing gaseous exchange. We have found it unnecessary to use this indicator if the catalyst is changed as outlined above. When the jar is not in use it is advisable to keep the catalyst dry in order to prolong its life.

Isolation of Organisms

Plating of tissue fluids or smearing of the tissues directly on the plate followed by streaking produces colonies of the pathogen if the medium has been prepared as described above. Details of the cell and colonial morphology of a variety of clostridia have already been described in detail in a separate article (Batty and Walker, 1965). What follows serves to supplement the earlier paper but for completeness and convenience the photographs from this paper are appended at the end of this chapter.

Clostridium welchii

Isolation of *Cl. welchii* presents little difficulty. The organism grows well on ordinary blood agar and is not strict regarding its tolerance to oxygen. Blood agar colonies of the organism are usually smooth, round and glistening, *c.* 3–4 mm in diameter and depending on the strains are surrounded by a zone of haemolysis (Fig. 2). The zone of haemolysis is a particularly good marker for selecting *Cl. welchii* from contaminants such as *Escherichia coli* on plates from intestinal contents. However, haemolysis can be extremely variable and haemolytic and non-haemolytic colonies of the same strain can often be found on the same plate. In addition strains of *Cl. welchii* which do not form *theta* toxin, produce only an incomplete zone of haemolysis or no haemolysis at all and can easily be overlooked if haemolysis is regarded as the sole criterion for selection. Usually, however, the conditions which cause the disease select the causative type. For example, colonies from plates prepared from normal intestinal contents are invariably those of *Cl. welchii* type A, but in the case of enteroxaemia due to *Cl. welchii* types B, C and D, large numbers of these are present in the

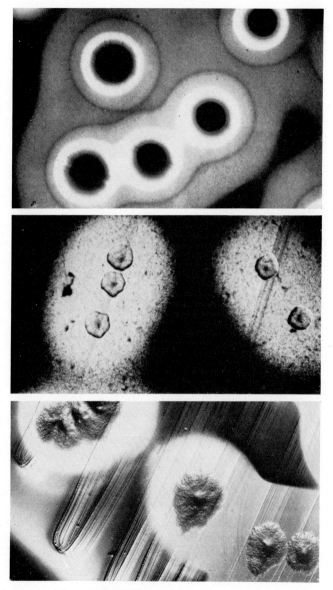

FIG. 2. Colonies of *Cl. welchii* type C after 24 h incubation on blood agar; note zones of haemolysis.

FIG. 3 Colonies of *Cl. chauvoei* after 24 h incubation on "*Cl. chauvoei* medium"; note wide zones of haemolysis.

FIG. 4. Colonies of *Cl. oedematiens* type A after 24 h incubation on blood agar; note wide zones of haemolysis.

gut and consequently a high proportion of the colonies grown on plates will be of these types. On lecithinase agar plates the colonies are surrounded by a zone of opacity (Fig. 1A) and this helps to differentiate the organism although similar zones may occur with colonies of *Cl. sordellii, Cl. oedematiens* and *B. cereus*. Colonies of the different types of *Cl. welchii* have identical morphology and it is impossible to differentiate on this basis. Rough colonies are occasionally present and can be recognized with experience.

Clostridium septicum

Cl. septicum grows readily on blood agar but because of its tendency to spread, the agar concentration is increased in order to obtain isolated colonies. Colonies of *Cl. septicum* are usually irregular with a rhizoidal edge but smooth, round colonies are produced by some strains. As mentioned earlier, although *Cl. septicum* causes specific disease in animals, it is commonly found as an invader in *post-mortem* tissues of animals which have died from other causes.

Clostridium chauvoei

Cl. chauvoei grows poorly on ordinary blood agar but good growth may be obtained by supplementing the medium with liver extract. Identification of the organism is greatly facilitated by incorporating sheep blood into the agar instead of horse blood because *Cl. chauvoei* produces a particularly effective sheep red cell haemolysin (Fig. 3). If the tissue is contaminated with *Cl. septicum* the agar concentration should be increased to 3 % (w/v) in order to obtain isolated colonies.

Clostridium sordellii

Cl. sordellii grows well on stiff agar and produces irregular colonies. These can easily be differentiated from *Cl. bifermentans* colonies which are smooth and round when grown on the same medium (Brooks and Epps, 1959). Colonies lose their translucent appearance and become white on ageing because of the production of spores.

Clostridium tetani

Cl. tetani again grows readily on blood agar and if 3 % agar is used separate colonies can be obtained. The typical drum-stick appearance is seen in films. *Cl. tetani* can usually be isolated from the lesions surrounding the rubber ring used for castrating lambs where the disease has developed.

Clostridium oedematiens type A

Unlike the other types of *Cl. oedematiens*, type A can grow on ordinary blood agar and produces irregular colonies with a rhizoidal edge surrounded by a large zone of haemolysis (Fig. 4). Types B, C and D are extremely demanding both in their nutritional requirements and in their tolerance of oxygen. In our hands consistent luxuriant growth on plates can only be obtained by meticulously following the methods of preparing and inoculating plates described below.

Clostridium oedematiens types B, C and D

The essential nature of L-cysteine for the successful cultivation of *Cl. oedematiens* type B on the surface of a solid medium (NP-Medium) and the obligation to protect this amino acid from loss by atmospheric oxidation, was demonstrated by Moore (1968). The composition of the medium is shown in Table 2 and the procedure adopted for its preparation has been

TABLE 2. The composition of NP—Medium used for plating cultures of *Cl. oedematiens* type B

"Peptone"*	10 g	Cysteine hydrochloride	100 mg
Yeast Extract†	5 g	Glutamine	50 mg
Proteolysed Liver‡	5 g	Dithiothreitol	100 mg
Glucose	10 g	Horse blood	100 ml
Agar	20 g	Salts solution§	5 ml
Glass-distilled water to 1000 ml; pH 7·6–7·8			

* Neopeptone (Difco); Bacto-Tryptone (Difco); Tryptone (Oxoid); Trypticase Soy Broth (Oxoid); Protein Hydrolysate (Burroughs Wellcome); each at equivalent total nitrogen concentration.
† Difco.
‡ Pabryn Laboratories, Greenford, Middlesex.
§ Salts solution contained (g/l); $MgSO_4.7H_2O$, 40; $MnSO_4.4H_2O$, 2; $FeCl_3$ (anhydrous), 0·4; and concentrated HCl, 0·5 ml.

described elsewhere (Moore, 1968). It is probable that almost any "peptone" preparation can be used as a source of peptides and amino acids, provided that the essential presence of the cysteine + dithiothreitol (DTT) mixture is recognized. Those preparations which have been found to be successful are: Protein Hydrolysate (Wellcome Reagents Ltd.), Bacto-Tryptone (Difco), Neopeptone (Difco), Tryptone (Oxoid), Trypticase Soy Broth (Oxoid) and Robertson's Cooked Meat Broth without meat particles. For the growth of some variants even acid hydrolysed casein can be used (CA-Medium).

The response of *Cl. oedematiens* on NP-Medium to increasing concentrations of L-cysteine in the presence of a fixed concentration of DTT,

Fig. 5. The effect of increasing concentrations of cysteine, in the presence of a fixed concentration of Dithiothreitol on the growth of toxinogenic and non-toxinogenic variants of *Cl. oedematiens* type B on NP-medium containing blood; Upper series: toxinogenic variant, Lower series: non-toxinogenic variant. Concentration of L-cysteine (μg/ml), right to left: 0, 0·1 1, 10, 100, 1000.

shown in Fig. 5, demonstrates clearly the importance of this amino acid for the luxuriant growth of this organism. DTT can be replaced by equi-molecular quantities of ascorbic acid or sodium sulphite but sodium thioglycollate often proves either less effective or ineffective. None of these compounds will replace L-cysteine itself.

The stimulatory effect of added L-tryptophan (Takarabe, 1960) on the growth of type B was demonstrated using liquid CA-Medium containing cysteine and thioglycollate. Results shown in Table 3 indicate that the

TABLE 3. The effects of L-tryptophan and carbohydrate energy source on the growth of toxinogenic and non-toxinogenic variants of *Clostridium oedematiens* type B in liquid CA medium (Moore, 1968) containing L-cysteine (100 μg/ml)

Type of variant	Carbon source	Concentration of added L-tryptophan μg/ml	Growth (24 h 37°) μg cell dry wt/ml
Toxinogenic	Fructose	0	150
		1000	950
	Glucose	0	320
		1000	600
Non-toxinogenic	Fructose	0	1400
		1000	1800
	Glucose	0	2000
		1000	2720

effect depends upon the strain used and the carbohydrate employed as carbon–energy source. However, since the effect is most marked for toxi-nogenic variants the inclusion of L-tryptophan at 1 mg/ml in either liquid or solidified media seems to be indicated. Non-toxinogenic variants grow more prolifically and are considerably less influenced by either carbon-energy source or L-tryptophan.

For some strains of *Cl. oedematiens* type B complete NP-Medium is apparently still inadequate. Such strains demand incubation of the medium in H_2-CO_2(95:5 v/v) mixture prior to inoculation. Provided that such pre-incubated plates are quickly streaked and re-established under anaerobic conditions good growth ensues. If there is a lag of more than *c.* 15 min before streaking subsequent growth is much reduced or even prevented. However, inclusion of sterile crystalline catalase (100 μg/ml) in the medium seems to obviate the need for this urgency, suggesting that peroxides are responsible for the inhibitory effect of medium which has not been pre-incubated. In order to cover all contingencies it would thus appear desirable to plate on basal medium containing blood, cysteine, DTT **and** catalase, or on pre-incubated medium if catalase is not available. The effects of

D

pre-incubation and catalase on growth of two strains of *Cl. oedematiens* type B are shown in Fig. 6.

Clostridium botulinum types C and D

Although the authors have no experience in isolating *Cl. botulinum* from pathogenic material, pure cultures of the organism produce luxuriant growth on the medium described above and are seen in the appended photographs. Other types of *Cl. botulinum*, associated with botulism in man, grow well on ordinary blood agar.

General Discussion

Most of the members of the genus *Clostridium*, particularly those species pathogenic for man and animals, display some degree of fastidiousness with regard to their requirements for growth apart from the obvious pre-requisite of anaerobiosis. *Cl. welchii*, for example, perhaps the least exacting of the pathogens, requires carbohydrate (glucose), organic nitrogen (amino acids) and a number of accessory factors such as the vitamin B-complex, before growth will take place. *Cl. oedematiens* type B, at the other extreme, requires in addition cysteine *per se* and a stricter control of anaerobic conditions by poising the medium at a low redox potential with cysteine, dithiothreitol and/or ascorbic acid and catalase or by pre-incubating the medium in H_2-CO_2(95:5 v/v) mixture. Growth is also markedly stimulated by L-tryptophan (1 mg/ml) and by glycyl-L-asparagine (Takarabe, 1960). Khairat (1966) drew attention to the desirability of including both yeast extract and proteolysed liver into media because of their synergistic effect on the growth of a very wide range of micro-organisms. Although one or other of these have been used in media for the cultivation of clostridia the combination has rarely, if ever, been employed. Many workers have stressed the importance of "fresh media" for the successful cultivation of "difficult" clostridia but the indication from this that cysteine *per se* could be involved does not appear to have been recognized. Cysteine is unique among the eighteen commonly occurring amino acids in the possession of a free thiol group. The latter is extremely susceptible to atmospheric oxidation, particularly at pH values above 7·0 in the presence of copper and ferrous ions (Elliott, 1930), conditions which clearly exist in media used for the culture of many microorganisms. Whilst it is appreciated that many of the growth factors of the vitamin B-complex are destroyed by light and/or heat and that procedures should be adopted to minimize their loss in culture media, there does not appear to be such a general awareness of the lability of cysteine to air and of the necessity to

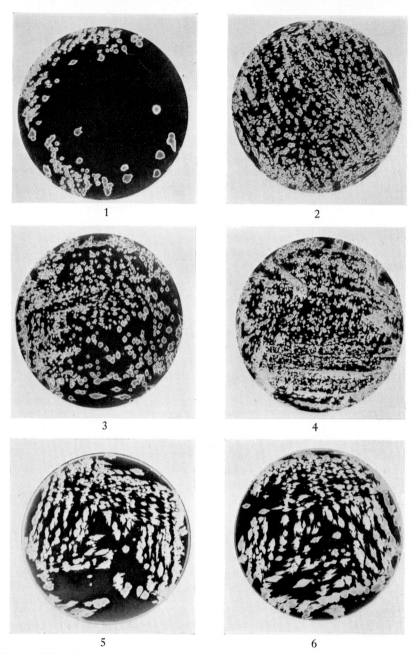

Fig. 6. The effect of pre-incubation of medium in H_2–CO_2, or supplementation with catalase on the growth of peroxide-sensitive and peroxide-insensitive strains of *Cl. oedematiens* on NP-medium. Growth of *Cl. oedematiens* type B strain A peroxide-sensitive, on NP-medium; 1. Medium not pre-incubated in H_2–CO_2, 2. Medium not pre-incubated in H_2–CO_2 but supplemented aseptically with crystalline catalase (100 μg/ml, final concentration), 3. Medium pre-incubated (2 h/37°), in H_2–CO_2, 4. Medium pre-incubated in H_2–CO_2 and supplemented with crystalline catalase. Growth of *Cl. oedematiens* type B strain B peroxide insensitive, on NP-medium; 5. Medium not pre-incubated, 6. Medium pre-incubated in H_2–CO_2.

protect it from oxidation. The use of dithiothreitol, which was introduced by Cleland (1964) for the protection of thiol groups, has helped enormously in the isolation and purification of compounds of biochemical interest containing this labile group.

A medium such as NP-medium which is adequate for the growth of the more fastidious species also supports luxuriant growth of less demanding ones. The use of a single medium for the isolation of all the pathogenic clostridia can thus be envisaged. It is unlikely, however, that routine diagnostic laboratories would be willing to prepare regularly such a medium as that described above for *Cl. oedematiens* type B. Since the majority of clostridia grow on blood agar, it would be convenient to use this medium for routine isolation whilst reserving more complex media for special culturing if and when required.

References

BATTY, I. & WALKER, P. D. (1963*a*). The differentation of *Clostridium septicum* and *Clostridium chauvoei* by the use of fluorescent labelled antibodies. *J. Path. Bact.*, **85**, 517.

BATTY, I. & WALKER, P. D. (1963*b*). Fluorescent labelled clostridial antisera as specific stains. *Null. Off. Int. Epiz.*, **59**, 1499.

BATTY, I. & WALKER, P. D. (1965). Colonial morphology and fluorescent labelled antibody staining in the identification of species of the genus *Clostridium*. *J. appl. Bact.*, **28**, 112.

BATTY, I., BUNTAIN, D. & WALKER, P. D. (1964). *Clostridium oedematiens*: A cause of sudden death in sheep, cattle and pigs. *Vet. Rec.*, **76**, 115.

BROOKS, M. E. & EPPS, H. B. G. (1959). Taxonomic studies of the genus *Clostridium*: *Clostridium bifermentans* and *Clostridium sordellii*. *J. gen. Microbiol.*, **21**, 144.

CLELAND, W. W. (1964). Dithiothreitol. A new protective reagent for SH groups. *Biochemistry*, **3**, 480.

ELLIOTT, K. A. C. (1930). XXXVII. On the catalysis of the oxidation of cysteine and thioglycollic acid by iron and copper. *Biochem. J.*, **24**, 310.

KHAIRAT, O. (1966). Efficient general-purpose culture medium for aerobes and anaerobes. *Can. J. Microbiol.*, **12**, 323.

MOORE, W. B. (1968). Solidified media suitable for the cultivation of *Clostridium novyi* Type B. *J. gen. Microbiol.*, **53**, 415.

TAKARABE, M. (1960). Studies on the nutritional requirements of some strains of *Clostridium novyi*. *Jap. J. Bact.*, **15**, 785.

WILLIS, A. T. (1960). *Anaerobic bacteriology in clinical medicine*. London: Butterworth & Co. (Publishers) Ltd.

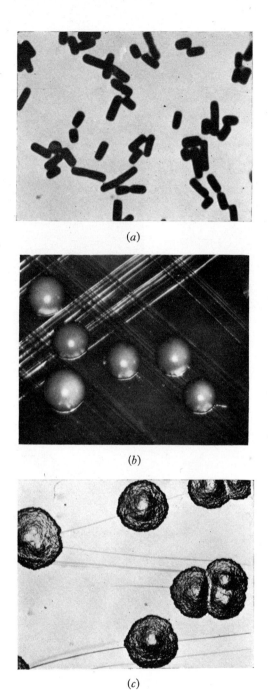

(a)

(b)

(c)

FIG. 1. Morphological appearances of *Cl. welchii* (a) from a surface colony 24 h old
(× 2000); (b) smooth surface colonies 24 h old (× 6); (c) rough surface colonies 24 h
old (× 6).

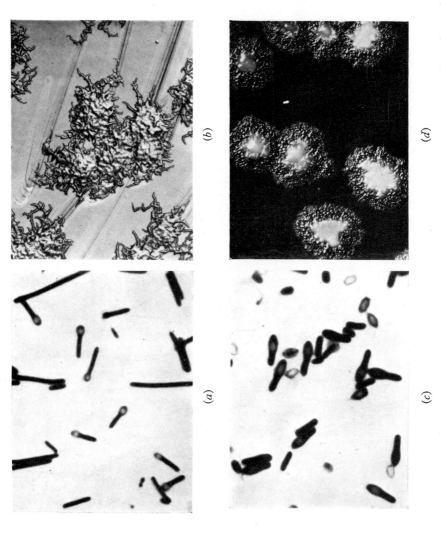

(a)

(b)

(c)

(d)

Fig. 2. Morphological appearances of *Cl. tetani* (*a*) from a surface colony 48 h old (×2000); (*b*) surface colony 48 h old (×6); and of *Cl. sporogenes* (*c*) from a surface colony 24 h old (×2000); (*d*) surface colonies 24 h old (×6).

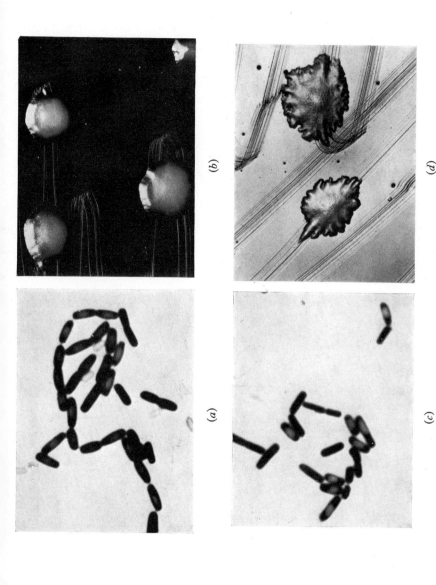

Fig. 3. Morphological appearances of *Cl. bifermentans* (*a*) from a surface colony 24 h old (×2000); (*b*) surface colonies 24 h old (×6); and of *Cl. sordellii* (*c*) from a surface colony 24 h old (×2000); (*d*) surface colonies 24 h old (×6).

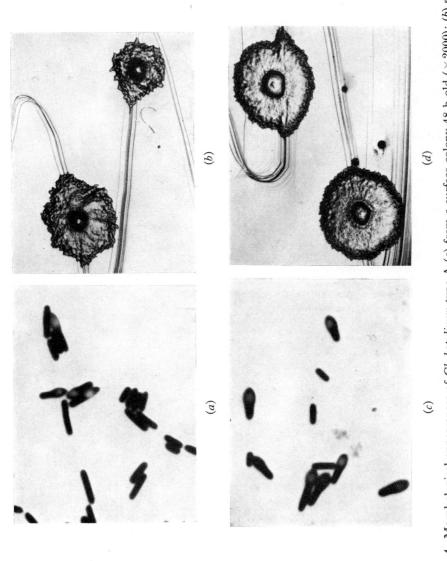

FIG. 4. Morphological appearances of *Cl. botulinum* type A (*a*) from a surface colony 48 h old (×2000); (*b*) surface colonies 48 h old (×6); and of type B (*c*) from a surface colony 48 h old (×2000); (*d*) surface colonies 48 h old (×6).

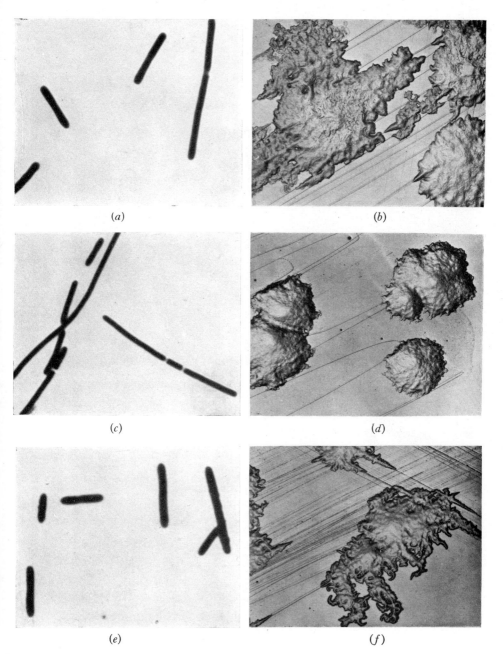

FIG. 5. Morphological appearances of *Cl. botulinum* type C (*a*) from a surface colony 48 h old (×2000); (*b*) surface colonies 48 h old (×6); of type D (*c*) from a surface colony 48 h old (×2000); (*d*) surface colonies 48 h old (×6); and of type E (*e*) from a surface colony 48 h old (×2000); (*f*) 48 h old (×6) surface colonies.

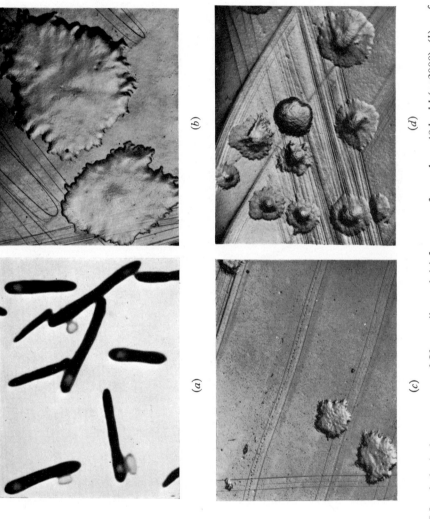

(a)

(b)

(c)

(d)

FIG. 6. Morphological appearances of *Cl. novyii* type A (*a*) from a surface colony 48 h old (×2000); (*b*) surface colonies 48 h old (×6); of type B (*c*) surface colonies 48 h old (×2000); and of type D (*d*) surface colonies 48 h old (×6).

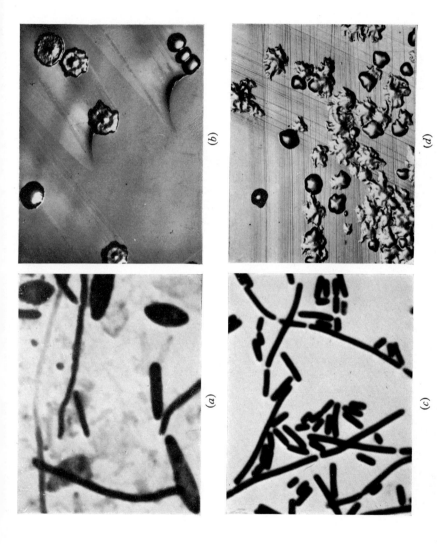

Fig. 7. Morphological appearances of *Cl. chauvoei* (*a*) from a surface colony 48 h old (×2000); (*b*) surface colonies rough and smooth 48 h old (×6); and of *Cl. septicum* (*c*) from a surface colony 24 h old (×2000); (*d*) surface colonies rough and smooth 24 h old (×6).

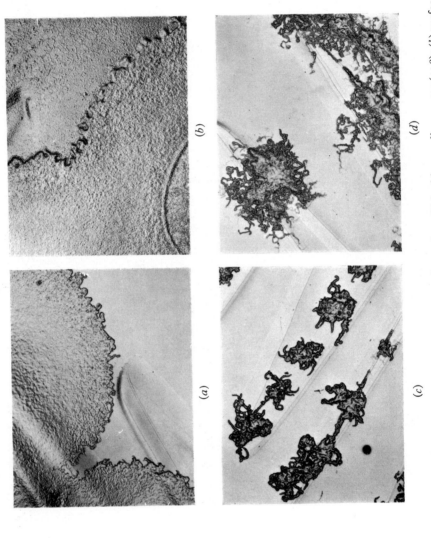

FIG. 8. Morphological appearances of *Cl. septicum* (*a*) a surface colony 24 h old on ordinary agar (×6); (*b*) surface colony 48 h old on ordinary agar (×6); (*c*) surface colonies on 24 h old stiff agar (×6); (*d*) surface colonies on stiff agar 48 h old (×6).

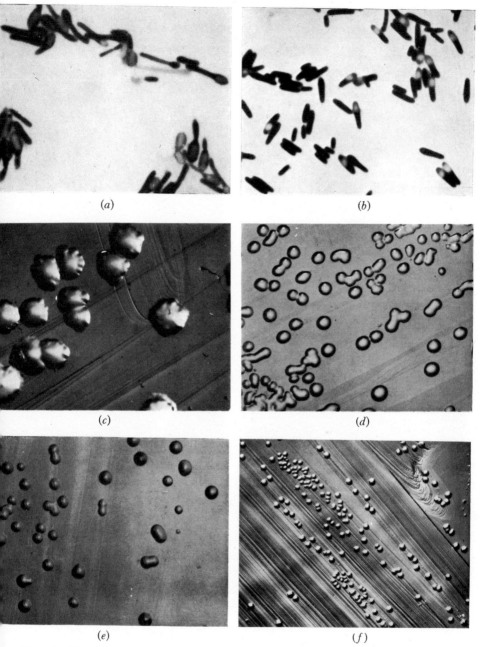

Fig. 9. Morphological appearances of *Cl. tertium* (*a*) from a surface colony 24 h old (×2000); (*c*) surface colonies anaerobic 24 h old (×6); (*e*) surface colonies aerobic 24 h old (×6); and of *Cl. histolyticum* (*b*) from a surface colony 24 h old (×2000); (*d*) surface colonies anaerobic 24 h old (×6); (*f*) surface colonies aerobic 24 h old (×6).

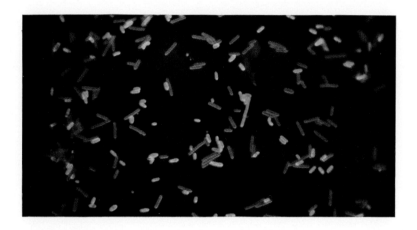

FIG. 10. A smear of a mixed *Cl. septicum* and *Cl. chauvoei* culture stained with a mixture of fluorescein isothiocyanate labelled *Cl. septicum* antiserum and lissamine rhodamine B.200 *Cl. chauvoei* labelled antiserum.

Isolation and Enumeration of *Clostridium welchii* from Food and Faeces

*R. G. A. Sutton, A. C. Ghosh and Betty C. Hobbs

*Food Hygiene Laboratory, Central Public Health Laboratory,
Colindale Avenue, London, England*

Although early workers described outbreaks of diarrhoea which they ascribed to *Clostridium welchii* (Klein, 1895; Andrewes, 1899; Simonds, 1915) little interest was taken in the role of *Cl. welchii* as a major cause of food poisoning until the report of Hobbs and her colleagues in 1953. They described a clearly defined form of food poisoning, with symptoms of diarrhoea and acute abdominal pain which usually began 8–14 h after ingesting a meat meal containing large numbers of viable *Cl. welchii* and persisted for 2–24 h. The patients had usually recovered within 24 h.

In England and Wales, *Cl. welchii* accounts for 2000–4000 individual cases of food poisoning annually; this represents 20–50% of all reported cases and the incidence of this form of food poisoning has shown little tendency to decline over the years (Vernon, 1969). In the U.S.A., also, attention is turning to the role of *Cl. welchii* as a food poisoning organism. During 1968, fifty-six large outbreaks of *Cl. welchii* food poisoning were recorded involving almost six thousand people; this is an increase of 100% on the figures recorded for 1967 (W.H.O., 1969). It is possible that much of this reported increase is due to a greater awareness of the role of *Cl. welchii* in food poisoning, and not entirely due to an increase in the prevalance of the disease.

The need for Public Health Laboratories to examine samples of food and faeces for *Cl. welchii* during the routine investigation of food poisoning outbreaks should be apparent; unfortunately this is still not carried out by some laboratories. The following methods have been found suitable for the purpose.

* Present address: Microbiology Department, School of Public Health and Tropical Medicine, University of Sydney, Sydney, Australia.

Examination of Faeces

Earlier workers (Klein, 1895; Simonds, 1915) isolated *Cl. welchii* from contaminated samples by heating at 80° in milk medium followed by incubation at 37°. Positive samples gave the characteristic stormy fermentation reaction in the milk medium.

With the publication of the work of Hobbs and her colleagues (1953), attention was focused on the presence of heat resistant *Cl. welchii* in faeces, and an enrichment culture after heating the faecal sample in a boiling water bath was used to detect the heat-resistant spores of the so-called "typical food poisoning strain" of *Cl. welchii*, type A.

Isolation of heat-resistant Cl. welchii

The method of Hobbs *et al.* (1953) is still the most widely used technique for the isolation of heat-resistant *Cl. welchii* from faeces. The method is as follows:

A small portion of faeces (about the size of a pea), is emulsified in a tube of cooked meat medium (CMM) which is heated in a boiling water bath for 60 min, incubated overnight at 37° and subcultured on to neomycin blood agar. Heat resistant *Cl. welchii* appears as a pure culture after anaerobic incubation at 37° for 24 h.

The use of nutrient broth or thioglycollate medium instead of CMM, and heating for 30, and not 60, min are variations that have been used by different workers, mainly for convenience in their particular laboratory, and not as a proposed improvement to the original method of Hobbs and her colleagues.

Using this technique, heat-resistant strains can be isolated from more than 90% of stool samples from persons with food poisoning due to heat resistant *Cl. welchii*, but from less than 10% of the healthy population (Sutton, 1966; Hobbs *et al.*, 1953). The technique, although only qualitative is therefore suitable for the diagnosis of food poisoning due to heat-resistant *Cl. welchii*.

Isolation of heat-sensitive Cl. welchii

Unlike the heat-resistant strain, heat-sensitive *Cl. welchii* is found in almost 100% of stool samples from healthy persons (Collee, Knowlden and Hobbs, 1961) and the mere isolation of this strain from stool samples is, therefore, of little significance. There is need for a test to distinguish faeces from persons with food poisoning from other faecal samples. Although this can (and should) be done by demonstrating a serological relationship between the strains isolated from food and faecal samples, it is not possible to serotype all strains, particularly the heat-sensitive strains, and only a few laboratories have a supply of antisera. The hospital and public health bacteriologist requires a simple test that will help him to assess the significance of the strains, and decide if it is worthwhile proceeding with a serological investigation.

Sutton (1969) carried out direct viable counts on faeces from healthy persons, persons with diarrhoea due to causes other than *Cl. welchii* (e.g. salmonellosis, dysentery) and from persons involved in 20 outbreaks of *Cl. welchii* food poisoning. Counts were carried out on tenfold dilutions of faecal suspensions in Ringer's solution, using the surface-dropping technique of Miles and Misra (1938), and neomycin blood agar (Sutton and Hobbs, 1968).

The results indicated that faeces from outbreaks of food poisoning due to heat-sensitive or heat-resistant *Cl. welchii* contain significantly larger numbers of *Cl. welchii* than those from healthy persons or persons with diarrhoea due to causes other than *Cl. welchii* food poisoning (Table 1).

TABLE 1. Viable counts of *Cl. welchii* in the faeces of persons with *Cl. welchii* food poisoning, persons with diarrhoea due to other causes and normal healthy persons

Range of Counts (*Cl. welchii*/g)	Number of specimens within range		
	Healthy Persons	Persons with Diarrhoea	*Cl. welchii* Food poisoning
$<5 \times 10^3$	20	35	25
5×10^3 to 5×10^5	26	31	38
$>5 \times 10^5$	4	15	189
Total number of samples	50	81	252
Median Count	$7 \cdot 5 \times 10^3$	$1 \cdot 5 \times 10^4$	$8 \cdot 5 \times 10^6$

Some form of viable count, therefore, is useful as a diagnostic test, particularly for outbreaks due to heat-sensitive strains.

Heat activation

Increased counts can be demonstrated before and after heating the faeces, although the count obtained before heating does not always give the total number of viable *Cl. welchii* present. In 6 out of 20 incidents investigated by Sutton (1969) the causative strain of *Cl. welchii* was present mainly as spores requiring heat shock before germination, and a low count (less than

$10^4/g$) before heating was misleading. This need for heat activation could be demonstrated only with non-haemolytic heat-resistant strains.

Table 2 gives the counts obtained for the individual faeces from an

TABLE 2.* Variation in viable counts of *Cl. welchii* before and after heating 8 faecal samples from an outbreak of food poisoning in which need for heat activation was demonstrated

Sample No.	Viable Count (*Cl. welchii*/g)		Ratio Spore : Unheated
	Before heating	After heating 80° for 10 min	
1	$5\cdot0 \times 10^3$	$1\cdot0 \times 10^6$	200:1
2	$5\cdot0 \times 10^3$	$1\cdot0 \times 10^7$	2000:1
3	$<5 \times 10^2$	$1\cdot5 \times 10^6$	>2000:1
4	$3\cdot5 \times 10^5$	$1\cdot0 \times 10^7$	30:1
5	$<5 \times 10^2$	$7\cdot0 \times 10^5$	>1000:1
6	$<5 \times 10^2$	$1\cdot0 \times 10^6$	>2000:1
7	$5\cdot0 \times 10^4$	$7\cdot0 \times 10^5$	14:1
8	$1\cdot5 \times 10^4$	$5\cdot0 \times 10^5$	30:1
Median	$5\cdot0 \times 10^3$	$1\cdot5 \times 10^6$	1000:1

* Taken from Sutton (1969).

outbreak in which heat activation could be clearly demonstrated. A mild heat treatment, such as 80° for 10 min, is therefore recommended if all outbreaks due to *Cl. welchii* are to be detected by direct culture.

Although an increased count after heating the faeces at 100° for 30 min can be demonstrated in stools from patients with food poisoning due to heat-resistant *Cl. welchii*, it is not recommended as a routine diagnostic procedure. It is useful, however, in distinguishing symptomless excretors of heat resistant *Cl. welchii* from active cases. *Cl. welchii* could be detected by direct culture in only 2 out of 22 symptomless excretors of heat-resistant *Cl. welchii* (Sutton, 1966).

Effect of collection time on viable counts

Although the viable count of *Cl. welchii* is high at the time of the acute food poisoning it begins to fall shortly afterwards. In 2–3 weeks most faecal samples will contain less than 10^5 *Cl. welchii*/g. Samples should be collected, therefore, within 3 days of the onset of symptoms. If there is a long delay, direct counts may be low and an enrichment technique, in CMM heated at 80° for 10 min, may be required. Hobbs *et al.* (1953) found that 19 out of 19 samples were positive for heat-resistant *Cl. welchii* at the time of an outbreak, 6 out of 10 were positive 13 days later, and only 1 out of 4 was positive 26 days after the onset of symptoms.

Rapid semiquantitative technique for routine examination of faeces

A Miles and Misra (1938) count, although relatively simple, is thought to be impractical for busy hospital or public health laboratories. The following simplified method has proved satisfactory; it was compared in parallel with the Miles and Misra method for six outbreaks of food poisoning and it is suggested as an alternative procedure:

1. A thick emulsion of faeces (1/5–1/10) is made in quarter-strength Ringer's solution.

2. With this emulsion a semi-quantitative direct count is carried out on neomycin blood agar using a calibrated loop in a manner similar to that now routinely used in many hospitals for urine culture. Ideally, this count should be done both without previous heating and after heating the emulsion to 80° for 10 min (spore count). If for any reason both counts cannot be made, that after heating is the more important. Plates are incubated anaerobically at 37° for 16–24 h (no longer). Using this technique it is possible to tell if *Cl. welchii* is present in the faeces in relatively small or large numbers.

3. The emulsion (1–2 ml) is inoculated into a tube of CMM and heated at 100° for 60 min. It is then incubated overnight at 37° and subcultured on to neomycin blood agar. This will detect the presence or absence of heat resistant spores of *Cl. welchii*.

Media routinely used for the isolation of salmonellae, shigellae and staphylococci can be inoculated using this emulsion, although this must be done before heating.

The procedure is easy to carry out and does not require media or materials not routinely found in hospital laboratories. It has been used by three Public Health Laboratory Service laboratories in a two-year trial, and all have found it to be satisfactory.

Although the direct culture (step 2) alone will suffice, and will detect outbreaks due to both heat-sensitive and heat-resistant strains, the use of a technique that will distinguish heat-sensitive from heat-resistant strains is suggested because: (a) knowledge of the heat-resistance of the causative strain may be helpful in finding the fault which gave rise to the outbreak; (b) much more information is needed regarding the relative number of outbreaks due to heat-sensitive strains, and (c) outbreaks have been recorded in which both heat-sensitive and heat-resistant strains were thought to be simultaneously involved (Sutton and Hobbs, 1968). Both strains may not be detected unless a direct culture and an enrichment culture after boiling are used, as the heat-sensitive strain may be present in the faeces in larger numbers and so mask the presence of the heat-resistant strain on direct culture.

E

Examination of Food Samples

A Miles and Misra (1938) counting technique on neomycin blood agar, using dilutions of the food homogenized in Ringer's solution can be used for viable counts of *Cl. welchii* in foods. Cooked samples of food from outbreaks will rarely contain spores, as they will have germinated. Therefore, the food sample should not be given a heat treatment before being cultured. Such factors as uneven distribution of the organism throughout the food, and a reduction in the number of organisms due to freezing of the sample before laboratory examination may result in a low count of *Cl. welchii* in the incriminated food; an enrichment technique, in CMM containing 100 μg/ml neomycin sulphate, should therefore supplement the direct viable count.

Identification of Cl. welchii

As a short routine procedure, colonies giving the typical appearance of *Cl. welchii* on blood agar may be further examined by Gram's method, and for lactose fermentation and inhibition of the lecithinase reaction by *Cl. welchii* type A antiserum on egg yolk medium. A more complete list of the properties of *Cl. welchii* can be found in any comprehensive textbook of bacteriology. Isolates of *Cl. welchii* should be serotyped in an effort to establish a serological relationship between strains isolated from food and those isolated from the individual faecal samples. Unfortunately, because sera are not available to all, only a limited number of laboratories are at present able to serotype *Cl. welchii*, type A.

Examination of samples not involved in food-poisoning outbreaks

Although the isolation and enumeration of *Cl. welchii* from foods is not a routine procedure in quality control laboratories, there is growing interest in the incidence of *Cl. welchii* in foodstuffs, and particularly in processed foods. *Cl. welchii* are usually present in such products but in small numbers only, and direct culture methods using serial dilutions of the food, and selective agar media are of little use, especially if the viable count of *Cl. welchii* is less than 10/g. Under these circumstances enrichment culture methods in liquid media must be used. Sutton and Hobbs (1969) used a most probable number (MPN) technique to estimate the number of *Cl. welchii* in a range of dehydrated foods with a median count for *Cl. welchii* of 25/100 g. Tubes of selective liquid media were inoculated with 3×10 g, 3×1 g and 3×0.1 g of the food, incubated overnight and subcultured on to neomycin blood agar. Results were reported as presence or absence of growth and the most probable number of *Cl. welchii* in the original sample was estimated from probability tables.

The Choice of a Suitable Medium

The choice of an isolation medium will depend on the preferences of the user and the purpose for which it is required. For the isolation of *Cl. welchii* from the mixed flora in food and faeces it is necessary not only to encourage the growth, but also to select *Cl. welchii* by suppressing the growth of other organisms and by the formation of easily recognized colonies. The most widely used media for this purpose are neomycin blood agar or one of the several variations of sulphite agar (Mossel *et al.*, 1956; Angelotti, Hall, Foter and Lewis, 1962; Marshall, Steenbergen and McClung, 1965). Neomycin blood agar is recommended for routine use because:

1. It shows haemolysis, which is useful when investigating outbreaks due to more than one strain of *Cl. welchii*, and in separating causative from indigenous strains.

2. The colonial morphology is consistent and an aid to identification.

3. It is readily available in public health and hospital laboratories.

4. It can be used for surface plating and is therefore suitable for use with the semi-quantitative technique suggested in this paper. The sulphite media, however, give their characteristic colonies only with a pour-plate method; and any quantitative technique using these media is therefore laborious when a large number of samples must be examined.

Of the available liquid media, cooked meat medium (made with fresh meat, and not commercially available dehydrated medium), Reinforced Clostridiul Medium (Hirsch and Grinsted, 1954) and thioglycollate medium are the most widely used. The addition of neomycin sulphate in a concentration of $100 \mu g/ml$ confers selectivity for clostridia. A comparison of the various agar and liquid media routinely used for the isolation of *Cl. welchii* is reported by Sutton and Hobbs (1969).

Media recommended for routine use

The three media routinely used in this laboratory for the investigation of outbreaks of *Cl. welchii* food poisoning are prepared as follows.

Neomycin blood agar

Three drops (0·06 ml) of a 1% solution of neomycin sulphate (Upjohn) are spread evenly over the surface of a 5% horse blood agar plate immediately before use. The 1% solution of neomycin sulphate will keep for at least six months if stored in the refrigerator.

Egg yolk medium

The lactose egg yolk medium of Willis and Hobbs (1958) is recommended. One half of the plate is spread with 2 drops of *Cl. welchii* type A antitoxin (Burroughs Wellcome) to inhibit the lecithinase (Nagler) reaction of *Cl. welchii*. This medium can also be used to determine lactose fermentation.

Cooked meat medium

This is prepared according to Cruickshank (1965) but using boneless veal instead of bullocks' hearts. The medium contains at least 50% meat particles.

Discussion

An assessment of the number of intestinal pathogens in both food and faeces is important for at least two reasons. It is unlikely that a foodstuff will give rise to clinical symptoms unless the number of *Cl. welchii* has risen to some millions/g of food. However, it is important to appreciate that a dried food ingredient even with a lower range of counts may introduce too many spores or even vegetative cells of *Cl. welchii* into processed foods if the count is greater than 100/g.

The stools of excreters may be harmless and normal when they are well formed and contain no more than the average 10^3 to 10^5 *Cl. welchii*/g. Many persons suffering from an acute attack of food poisoning will have a significantly raised count of *Cl. welchii* which is a useful diagnostic factor in addition to or in the absence of serological identification of the strains. Furthermore, all fluid faeces with abnormal counts of pathogens are potentially dangerous in any environment whether it be that of the toilets and the surrounding accommodation near kitchens, in hospital wards and sluice rooms, or in homes and shops. Between the two extremes of normality and abnormality there may well be varying degrees of fluidity of stool and raised counts when a numerical index could be a useful diagnostic guide. With an intestinal pathogen as ubiquitous as *Cl. welchii* enumeration as a guide to significance is almost essential.

Surveys of work on heat resistance and other characteristics of *Cl. welchii* are given by Hobbs and Sutton (1968*a*, *b*) and Hobbs (1969).

References

ANDREWES, F. W. (1899). On an outbreak of diarrhoea in the wards of St. Bartholomew's hospital, probably caused by infection of rice pudding with *Bacillus enteritidis sporogenes*. *Lancet* **i**, 8.

ANGELOTTI, R., HALL, H. E., FOTER, M. J. & LEWIS, K. H. (1962). Quantitation of *Clostridium perfringens* in foods. *Appl. Microbiol.*, **10**, 193.

COLLEE, J. G., KNOWLDEN, J. A. & HOBBS, B. C. (1961). Studies on the growth, sporulation and carriage of *Clostridium welchii* with special reference to food-poisoning strains. *J. appl. Bact.,* **24,** 326.

CRUICKSHANK, R. (1965). *Medical Microbiology,* 11th Ed., p. 775. London and Edinburgh: E. & S. Livingstone.

HIRSCH, A. & GRINSTED, E. (1954). Methods for the growth and enumeration of anaerobic spore-formers, from cheese, with observations on the effect of nisin. *J. Dairy Res.,* **21,** 101.

HOBBS, B. C. (1969). *Clostridium perfringens* and *Bacillus cereus* infections. *Food Borne Infections and Intoxications* (ed. H. Reimann). New York and London: Academic Press, p. 131.

HOBBS, B. C. & SUTTON, R. G. A. (1968*a*). *Clostridium perfringens* food poisoning. *Annls. Inst. Pasteur Lille,* **19,** 29

HOBBS, B. C. & SUTTON, R. G. A. (1968*b*). Characteristics of spores of *Clostridium welchii* in relation to food hygiene. In: *The anaerobic bacteria*: proceedings of an international workshop, 1967; (ed. V. Fredette) p. 51, Institute of Microbiology and Hygiene of Montreal University.

HOBBS, B. C., SMITH, M. E., OAKLEY, C. L., WARRACK, G. H. & CRUICKSHANK, J. C. (1953). *Clostridium welchii* food poisoning. *J. Hyg., Camb.,* **51,** 75.

KLEIN, E. (1895). Ueber einen pathogen anaeroben Darmbacillus, *Bacillus enteritidis sporogenes. Zentbl. Bakt. Parasit Kde* Abt 1, Orig., **18,** 737.

MARSHALL, R. S., STEENBERGEN, J. P. & McCLUNG, L. S. (1965), Rapid technique for enumeration of *Clostridium perfringens. Appl. Microbiol.,* **13.** 559.

MILES, A. A. & MISRA, S. S. (1938). The estimation of the bactericidal power of blood. *J. Hyg., Camb.,* **38,** 732.

MOSSEL, D. A. A., BRUIN, A. S. DE, DIEPEN, H. M. J. VAN, VENDRIG, C. M. A. & ZOUTEWELLE, G. (1956). The enumeration of anaerobic bacteria, and of *Clostridium* species in particular, in foods. *J. appl. Bact.,* **19,** 142.

SIMONDS, J. P. (1915). Studies in *Bacillus welchii* with special reference to classification and to its relation to diarrhoea. Monograph of Rockefeller Institute for Medical Research, No. 5.

SUTTON, R. G. A. (1966). Enumeration of *Clostridium welchii* in the faeces of varying sections of the human population. *J. Hyg., Camb.,* **64,** 367.

SUTTON, R. G. A. (1969). The pathogenesis and epidemiology of *Clostridium welchii* food poisoning. Ph.D. thesis, Univ. of London.

SUTTON, R. G. A. & HOBBS, B. C. (1968). Food poisoning caused by heat-sensitive *Cl. welchii.* A report of five recent outbreaks. *J. Hyg., Camb.,* **66,** 135.

SUTTON, R. G. A. & HOBBS, B. C. (1969). The enumeration of *Clostridium perfringens* in dried foods and feeding stuffs. In: *The Microbiology of Dried Foods: Proc. VI. Int. Symp. Food Microbiol.* Bilthoven. 1968, p. 243.

VERNON, E. (1969). Food Poisoning and Salmonella infections in England and Wales 1967. *Publ. Hlth., Lond.,* **83,** 205.

W. H. O. (1969). Surveillance Summary of Foodborne Disease Outbreaks, U.S.A. 1968. *Wkly. epidem. Rec.,* W. H. O., **44,** 349.

WILLIS, A. T. & HOBBS, G. (1958). A medium for identification of clostridia producing opalescence in egg yolk emulsions. *J. Path. Bact.,* **75,** 299.

The Use of Differential Reinforced Clostridial Medium for the Isolation and Enumeration of Clostridia from Food

BARBARA FREAME AND B. W. F. FITZPATRICK

Unilever Research Laboratory, Colworth House, Sharnbrook, Bedford, England

Many of the selective media that are routinely used to enumerate clostridia in foods were designed primarily for the isolation of *Clostridium perfringens* (Angelotti, Hall, Foter and Lewis, 1962; Marshall, Steenbergen and McClung, 1965; Green and Litsky, 1966; Nakamura and Kelly, 1968; Hall, Witzeman and Janes, 1969; Mead, 1969). These media inhibit many other clostridia (Gibbs and Freame, 1965), it is therefore preferable to use a non-selective medium for routine and quality control counts of the diverse clostridia which may occur in most foods, coupled with a most probable number (MPN) procedure, which will permit low numbers of bacteria to be enumerated. Such a procedure was suggested by Gibbs and Freame (1965), who introduced Differential Reinforced Clostridial Medium (DRCM) for the detection and enumeration of clostridia in foods. DRCM is based on the Reinforced Clostridial Medium of Hirsch and Grinsted (1954) and Gibbs and Hirsch (1956); it is a rich medium that supports good growth of most clostridia, but also permits growth of other anaerobes and facultative anaerobes. In order to detect growth of clostridia in RCM, sodium sulphite and iron citrate were added. The sulphite is reduced by clostridia to form sulphide which, when combined to give iron sulphide, blackens the medium. This reaction was first used in Wilson and Blair's (1924) medium, and there have been many subsequent modifications described, including those by Mossel *et al.* (1956), Mossel (1959), Angelotti *et al.* (1962) and Mead (1969). It is important to realize that the blackening of DRCM is not specific for clostridia, but can be caused by other bacteria such as salmonellae, *Proteus* spp, bacteroides and some strains of *Escherichia coli*. However, as these organisms are not spore formers pasteurization or treatment with ethyl alcohol (Johnston, Harmon and Kautter, 1964) will eliminate them from samples.

Composition and preparation of DRCM

The basal medium contains: peptone (Evans), 10 g; Lab Lemco (Oxoid), 10 g; hydrated sodium acetate, 5 g; yeast extract (Difco), 1·5 g; soluble starch, 1 g; glucose, 1 g; L-cysteine HCl, 0·5g; distilled water, 1000 ml; pH 7·1–7·2. To prepare the basal medium add the peptone, Lab Lemco, sodium acetate and yeast extract to 800 ml of distilled water. A starch solution is prepared in the other 200 ml of water by making a cold paste in a little water, boiling the rest and then stirring it into the paste. The solutions are mixed and steamed for 30 min to dissolve the ingredients. After steaming, the glucose and cysteine are added and the pH is adjusted (7·1–7·2) with 10N-NaOH. The medium is filtered hot through paper pulp, dispensed in 25 ml amounts in screwcapped McCartney bottles (1 oz) and sterilized by autoclaving (121°/15 min). Immediately before use, sodium sulphite and ferric citrate are added to give final concentrations of 0·04 and 0·07% (w/v) (anhydrous salts) respectively. The sodium sulphite (anhydrous) is prepared as a 4% (w/v) solution and the ferric citrate (scales) as a 7% (w/v) solution; it is necessary to heat the ferric citrate for about 5 min to dissolve. Both solutions are sterilized by filtration and can be stored at 3–5° in fully filled McCartney bottles for at least 14 days. On the day that the DRCM is required, the basal medium is steamed and cooled; equal volumes of the sodium sulphite and ferric citrate solutions are mixed together and 0·5 ml of the mixture added aseptically to each 25 ml amount of DRCM.

Enumeration of Clostridia from Food using DRCM

At least 10 g of food is macerated in sufficient 0·1% peptone water to give a 0·2 or 0·1 (w/v) dilution of the food. Decimal dilutions of the macerate are prepared in the same diluent and 1 ml amounts of the original macerate plus at least the next 2 decimal dilutions transferred to bottles of DRCM for MPN counts. Inoculate either 3 or 5 tubes with each dilution depending on the accuracy required (Taylor, 1962).

Presumptive clostridial count (vegetative forms plus spores)

Bottles are incubated at 30° for at least 7 days; most positives appear within 3 to 4 days, but slow growing or slow germinating spores may not cause blackening of the medium until later. The number of bottles with blackened contents at each dilution are noted and the count obtained by reference to MPN tables (Demeter, Saur and Miller, 1933; Hoskins, 1934). This gives the **presumptive clostridial** count (vegetative cells and spores) only, since other organisms capable of blackening DRCM may be present. The contents of the bottles must normally be tested further to confirm that clostridia are present. However, this is necessary only if the total presumptive clostridial count is significantly higher than the corresponding spore count on the same macerate (see *Spore count*, below), because blackening in the spore count is **not** affected by non-clostridial forms. To confirm the presence of clostridia either pasteurize or alcohol treat (see *Spore count*, below) the contents of all bottles containing blackened DRCM from dilutions higher than the spore count. This procedure will destroy any non-sporing forms which can cause blackening and which may not be clostridia, whilst leaving viable the clostridia which will usually have formed spores. Each of the treated bottles is subcultured (1 ml) into a fresh bottle of DRCM and incubated at 30° for 7 days. Cultures that blacken the DRCM are scored as clostridia, and by reference back to the MPN tables the **total clostridial count** is obtained.

Spore count

A count of ungerminated clostridia spores is obtained by either pasteurizing or alcohol treating the food suspension so as to destroy vegetative microorganisms. For routine purposes pasteurization is preferred as it has the additional benefit of heat shocking any spores that are present thus allowing better germination (Barnes, Despaul and Ingram, 1963; Ingram, 1963; Duncan and Strong, 1968).

Pasteurization

Approximately 10 ml of the 0·2 or 0·1 dilution of the food macerate is sealed in a thin-walled 10 ml glass ampoule and heated by submerging in a 75° water bath for 30 min. If bottles are used instead of ampoules, the heating time and temperature will need adjusting to allow for heat penetration. For example, heat McCartney bottles at 75° for 40 min: ensure that the contents of the bottles are completely submerged in the heating bath, and that, when subculturing from these, there is no carry-over of viable vegetative cells which may have survived on the necks of the bottles (Gibbs and Hirsch, 1956).

Alcohol treatment

A portion of the food macerate is mixed with an equal amount of ethyl alcohol (rectified spirit) and left at room temperature for 1h with occasional shaking.

The MPN counts are obtained after incubation of cultures at 30° for at

F

least 7 days as before and the **clostridial spore count** is obtained by reference to MPN tables. There is usually little change in count after the first few days' incubation, however, positive tubes occasionally appear even after 4 weeks' incubation.

Isolation of Pure Cultures

Pure cultures can be obtained from DRCM tubes by plating on an agar medium such as blood agar and incubating in an anaerobic jar, or by deep agar tube methods which do not require incubation in O_2 free atmospheres. The use of DRCM agar (prepared as for DRCM with the addition of 1.5% (w/v) agar) in deep agar tubes is particularly useful for the initial isolation of clostridia from mixed cultures, as black colonies can easily be selected from any white (non-clostridial) colonies present. For this method we use open-ended tubes (c. 120 mm $\times$ 13 mm) which are closed at the bottom by rubber bungs and at the top by cotton wool plugs. Known (1 ml) amounts of decimal dilutions of the blackened DRCM culture are inoculated into the tubes and DRCM agar is poured to nearly fill the tubes. The tubes are then incubated at $30°$ until discrete black colonies can be picked. This is done by removing the rubber bung from the tube and pushing the cotton wool plug down to force the agar out of the bottom of the tube. When the colony required has reached the end of the tube the extruding agar is aseptically trimmed off and the colony picked, by means of a Pasteur pipette, into DRCM. The subsequent culture is checked for purity by plating out anaerobically, and if necessary can be re-passaged through deep agar tubes. Identification of the pure cultures can be made by using procedures such as those of Spray (1936), Willis and Hobbs (1959), or Smith and Holdeman (1968).

Discussion

We have found DRCM useful for the enumeration of clostridial spores in Pasteurized samples from foods. Counts from untreated food samples (the presumptive total clostridial vegetative cells + spore count) are, however, difficult to interpret, as vegetative organisms other than clostridia can produce blackening in DRCM. The subsequent treatment of blackened DRCM cultures from a presumptive total clostridial count presupposes that these cultures, if they are clostridia, will have formed some spores. Although most clostridia can sporulate in DRCM, not all species or strains do so. For instance, it is well known that *Cl. perfringens* is a poor spore former in culture media and specially formulated media are required to induce sporulation (Ellner, 1956; Hall, Angelotti, Lewis and Foter, 1963;

Hobbs, 1965; Duncan and Strong, 1968; Nishida, Seo and Nakagawa, 1969). If germinated spores or vegetative cells of *Cl. perfringens* are suspected, it would be advisable to use a medium specially selective for *Cl. perfringens* (see introduction), in addition to pasteurizing or alcohol treating the blackened cultures. Ingram (1963) pointed out that interpretation of total clostridial counts may be difficult because many spores will only germinate after a heat shock; counts on unheated samples may therefore not detect all the spores that are present. If the total (non-pasteurized) count is significantly higher than the corresponding spore count it could indicate that germination and growth of clostridia has occurred, so that the food sample contains many vegetative clostridial forms. In fact, this situation is rarely found in practice; the clostridia in unspoiled foods occur mostly in the spore form.

Although most clostridia blacken DRCM readily, some strains of *Cl. tyrobutyricum, Cl. saccharobutyricum, Cl. septicum* and *Cl. tertium* give a variable reaction unless the DRCM is overlayered with a Vaseline seal immediately after inoculation. *Cl. oedematiens* types B and D may not react in DRCM even when overlayered with Vaseline seal. However, these are fastidious clostridia which normally require special isolation procedures (Smith and Holdeman, 1968); the more commonly occurring clostridia grow well in DRCM.

Gibbs and Freame (1965) listed the advantages of using DRCM as: (a) good recovery of low numbers of clostridia; (b) no need to use anaerobic jars; (c) incubation can be prolonged without dehydration of the medium; (d) gas formation is no problem as it can be in agar cultures; (e) cultures can be examined daily without disturbing the growth conditions, and (f) *Cl. botulinum* is able to produce good titres of toxin in DRCM: toxicity can be tested directly by animal inoculation of the supernatant of centrifuged DRCM cultures. The disadvantages were listed as: (a) use of the MPN count, which is less accurate and more tedious in preparation than a plate count; (b) the need to make aseptic additions of the sulphite/citrate solution to each bottle of basal medium immediately before use, and (c) the growth and blackening in non-pasteurized counts by some bacteria other than the clostridia.

Gibbs and Freame (1965) found that polymyxin (70 units/ml) inhibited many of the non-clostridial forms capable of blackening DRCM. More effective selective agents would certainly increase the value of this medium; however, in their absence, the methods described in this paper provide a useful regime for enumerating clostridia, in which the limitations are few and well defined.

References

ANGELOTTI, R., HALL, H. E., FOTER, M. J. & LEWIS, K. H. (1962). Quantitation of *Clostridium perfringens* in foods. *Appl. Microbiol.*, **10**, 193.

BARNES, E. M., DESPAUL, J. E. & INGRAM, M. (1963). The behaviour of a food-poisoning strain of *Clostridium welchii* in beef. *J. appl. Bact.*, **26**, 415.

DEMETER, K. J., SAUR, F. & MILLER, M. (1933). Vergleichende Untersuchungen über verschiedene Methodene zur Coli-aerogenes-Tittersbestimmung in Milch. *Milchw. Forsch.*, **15**, 265.

DUNCAN, C. L. & STRONG, D. H. (1968). Improved medium for sporulation of *Clostridium perfringens*. *Appl. Microbiol.*, **16**, 82.

ELLNER, P. D. (1956). A medium promoting rapid quantitative sporulation in *Clostridium perfringens*. *J. Bact.*, **71**, 495.

GIBBS, B. M. & FREAME, B. (1965). Methods for the recovery of clostridia from foods. *J. appl. Bact.*, **28**, 95.

GIBBS, B. M. & HIRSCH, A. (1956). Spore formation by *Clostridium species* in an artificial medium. *J. appl. Bact.*, **19**, 129.

GREEN, J. H. & LITSKY, W. (1966). A new medium and "mimic" MPN method for *Clostridium perfringens* isolation and enumeration. *J. Food Sci.*, **31**, 610.

HALL, H. E., ANGELOTTI, R. LEWIS, K. H. & FOTER, M. J. (1963). Characteristics of *Clostridium perfringens* strains associated with food and foodborne disease. *J. Bact.*, **85**, 1094.

HALL, W. M., WITZEMAN, J. S. & JANES, R. (1969). The detection and enumeration of *Clostridium perfringens* in foods. *J. Fd Sci.*, **34**, 212.

HIRSCH, A. & GRINSTED, E. (1954). Methods for the growth and enumeration of anaerobic spore-formers from cheese, with observations on the effect of nisin. *J. Dairy Res.*, **21**, 101.

HOBBS, B. C. (1965). *Clostridium welchii* as a food-poisoning organism. *J. appl. Bact.*, **28**, 74.

HOSKINS, J. K. (1934). Most probable numbers for evaluation of coli-aerogenes tests by fermentation tube method. *Publ. Hlth. Rep. Wash.*, **49**, 393.

INGRAM, M. (1963). Difficulties in counting viable clostridia in foods *R. C. 1st. sup. Sanit.*, **26**, 330.

JOHNSTON, R., HARMON, S. & KAUTTER, D. (1964). Method to facilitate the isolation of *Clostridium botulinum* type E. *J. Bact.*, **88**, 1521.

MARSHALL, R. S., STEENBERGEN, J. F. & McCLUNG, L. S. (1965). Rapid technique for the enumeration of *Clostridium perfringens*. *Appl. Microbiol.*, **13**, 559.

MEAD, G. C. (1969). The use of the sulphite-containing media in the isolation of *Clostridium welchii*. *J. appl. Bact.*, **32**, 358.

MOSSEL, D. A. A. (1959). Enumeration of sulphite-reducing clostridia occurring in foods. *J. Sci. Fd Agric.*, **10**, 662.

MOSSEL, D. A. A., DE BRUIN, A. S., VAN DIEPEN, H. M. J., VENDRIG, C. M. A. & ZOUTWELLE, G. (1956). The enumeration of anaerobic bacteria and of *Clostridium* species in particular, in foods. *J. appl. Bact.*, **19**, 142.

NAKAMURA, M. & KELLY, K. D. (1968). *Clostridium perfringens* in dehydrated soups and sauces. *J. Fd Sci.*, **33**, 424.

NISHIDA, S., SEO, N. & NAKAGAWA, M. (1969). Sporulation, heat resistance, and biological properties of *Clostridium perfringens*. *Appl. Microbiol.*, **17**, 303.

SPRAY, R. S. (1936). Semi-solid media for cultivation and identification of the sporulating anaerobes. *J. Bact.*, **32**, 135.

SMITH, L. DS. & HOLDEMAN, L. V. (1968). *The Pathogenic anaerobic bacteria,* Springfield: Charles C. Thomas.

TAYLOR, J. (1962). The estimation of numbers of bacteria by tenfold dilution series. *J. appl. Bact.*, **25**, 54.

WILLIS, A. T. & HOBBS, G. (1959). Some new media for the isolation and identification of clostridia. *J. Path. Bact.*, **77**, 511.

WILSON, W. J. & BLAIR, E. M. (1924). The application of a sulphite-glucose-iron-agar medium to the quantitative estimation of *B. welchii* and other reducing bacteria in water supplies. *J. Path. Bact.*, **27**, 119.

The Isolation of Soil Clostridia

F. A. Skinner

*Soil Microbiology Department, Rothamsted Experimental Station,
Harpenden, Hertfordshire, England*

Many species of clostridia can be isolated from soils (Breed, Murray and Smith, 1957) by methods developed for their isolation from foodstuffs and pathological materials (Willis, 1964). Clostridia in soil are readily counted with the Differential Reinforced Clostridial Medium (DRCM) of Gibbs and Freame (1965), and they can be isolated easily from positive (black) cultures. The corresponding agar medium (DRCA) can be used to purify isolates, or to isolate clostridia directly from dilutions of soil suspension. Reinforced Clostridial Agar (RCA) (see Hirsch and Grinsted, 1954) is not suitable for isolating soil clostridia because it is not sufficiently selective to inhibit growth of the facultative anaerobes that usually greatly outnumber the clostridia in soil.

Two groups of clostridia are especially interesting to soil microbiologists, the strains of *Clostridium pasteurianum* and related species that are able to fix gaseous nitrogen, and those species, such as *Cl. cellobioparum*, that decompose cellulose. Only these two groups are considered below.

Nitrogen-fixing Clostridia

The nitrogen-fixing, spore-forming anaerobe reported by Winogradsky (1893), and subsequently named *Clostridium pasteurianum* (Winogradsky, 1895), actively ferments sugars and produces acids (including butyric) and gas (a mixture of hydrogen and carbon dioxide). Some strains of *Cl. butyricum*, which closely resembles *Cl. pasteurianum*, and of other clostridia, have also been shown to fix gaseous nitrogen (Rosenblum and Wilson, 1949). Indeed, nitrogen fixation is doubtless a property shared by many strains of the strictly anaerobic butyric acid bacteria. These organisms grow well on media currently used for clostridia but it is clearly advantageous to isolate them directly on selective media containing no combined nitrogen.

Three methods have been used to isolate nitrogen-fixing clostridia from soil: enrichment culture followed by purification of isolates on solid medium, direct plating of soil dilutions on solid medium, and inoculation of liquid cultures with soil dilutions followed by purification on solid medium.

Enrichment is a convenient method when the need is to isolate a single strain that actively fixes nitrogen, but it tends to select rapidly growing strains, a feature that may not always be an advantage. The other two methods have been used primarily for counting nitrogen-fixers but they also enable slowly growing forms to be isolated, The usefulness of these methods depends greatly on composition of the media and, especially, on the content of trace elements and organic growth factors.

Isolation after enrichment

Winogradsky (1895) recommended the following procedure for isolating *Cl. pasteurianum*. A flask or gas-washing bottle containing sterile nitrogen-free mineral salts solution (Table 1) with glucose and chalk, is inoculated

TABLE 1. Mineral composition of different liquid media that have been used for growing nitrogen-fixing clostridia

Mineral salts added	Constituents of media (g/l)				
	A*	B	C	D	E
K_2HPO_4	1·0	0·5	0·25	0·5	0·8
KH_2PO_4	—	0·5†	0·75	0·5	0·2
$MgSO_4.7H_2O$	0·5	0·1	0·125	0·5	0·2
NaCl	0·01-0·02	0·1	0·125	0·5	0·2
$FeSO_4.7H_2O$	0·01-0·02	0·01	—	0·01	—
$Fe_2(SO_4)_3$	—	—	0·0025	—	0·01
$MnSO_4.7H_2O$	0·01-0·02	0·01	0·0025	0·01	—
$CaCO_3$	trace	4–5	—	—	—
$CaSO_4$	—	—	—	—	0·1

* A, Winogradsky (1895). B, Jensen and Spencer (1947). C, Augier (1957). Trace elements included. D, Emtsev (1962). Trace elements included. E, Burk (1930).
† Sometimes omitted.

with a little fresh soil and incubated at room temperature (18–20°) with a slow current of nitrogen bubbling through it. Fermentation is usually active by the third day. Three to five successive sub-cultures are similarly made with the same medium to eliminate most contaminants. Finally, a sporulating culture is pasteurized at 80° for 15 min, and inoculated on sterilized slices of potato, which are then incubated anaerobically. Originally, the inoculated slices were placed in a desiccator which was then evacuated. Any remaining traces of oxygen were absorbed by alkaline pyrogallol, freshly prepared *in situ* by tipping the vessel to mix the reagents after evacuation. Colonies are removed from the potato slices and the isolates purified by re-streaking on potato. Strains can now be purified by streaking on nitrogen-free agar and incubating the cultures in nitrogen without using alkaline pyrogallol (see below).

Similar enrichment methods, with minor variations, have been used by other workers (e.g. Jensen and Spencer, 1947).

Growth from small inocula

Winogradsky noticed that small inocula introduced into strictly nitrogen-free media sometimes failed to develop but that such failures could be prevented by adding small amounts of combined nitrogen, usually as ammonium sulphate, to the medium. These traces (e.g. 0·7 mg of combined nitrogen/g of glucose) had no effect on the final amount of nitrogen fixed but served only to "prime" the fermentation in nitrogen-free medium. Nitrogen fixation decreased with increasing amounts of ammonium sulphate and stopped when there were 5·8 mg of combined nitrogen/g of glucose. Winogradsky was concerned only with the amount, and not the nature, of the combined nitrogen, and he was unaware of the growth factors that are now known to be needed, and that were probably present in some of the nitrogen-containing supplements used (e.g. 1 ml of 0·2% peptone solution in 45 ml of nitrogen-free medium). Jensen and Spencer (1947) also found difficulty in obtaining growth from small inocula in a modified Winogradsky medium (Table 1), but they initiated growth reliably by adding potato extract.

A medium that will permit growth from very small inocula, even from single cells or spores, is especially needed for counting nitrogen-fixing anaerobes by MPN methods, or for isolating uncommon strains. It is convenient to discuss attempts to improve media by including different substances before describing methods suitable for isolating these organisms without previous enrichment.

Inorganic constituents of nitrogen-free media

Basal mineral salts

The basal nitrogen-free medium used by Winogradsky is given in Table 1: the precise composition differs slightly in different publications (Winogradsky, 1895, 1902; Waksman, 1927). The compositions of other media for the nitrogen-fixing clostridia are also presented in Table 1, which includes the medium devised by Burk (1930) for growing *Azotobacter* spp, but which can be made suitable for clostridia.

All the media listed contain potassium hydrogen orthophosphate as K_2HPO_4 alone, or mixed with KH_2PO_4 in different proportions. However, as the pH value is usually adjusted to *c*. 7·0, the exact type of orthophosphate supplied is unimportant. The inclusion of NaCl seems to depend more on personal preference than on any demonstrated need for this salt.

Iron

Iron was included in these media (see Table 1) long before Carnahan and Castle (1958) showed it to be necessary for nitrogen fixation. Growth of their nitrogen-fixing cultures was limited with less than 10 mg of $FeSO_4/l$, and was greatest with 100 mg of $FeSO_4/l$. On this showing, rather too little iron has been included in some media. Either ferrous or ferric salts can be used.

Molybdenum

The need to add molybdenum to cultures of nitrogen-fixing anaerobes (9 strains of *Cl. butyricum* and one of *Cl. acetobutylicum*) was established by Jensen and Spencer (1947). Though small amounts of nitrogen were fixed in the basal medium (Table 1) which was supplemented with potato extract (see below), fixation was increased 3- to 6-fold by adding traces of sodium molybdate. Enough molybdate to give 0·01 p/m of molybdenum was about optimal. Some strains (of *Cl. butyricum*) could use vanadium instead of molybdenum though it was less efficient. Molybdenum had little or no stimulating action on sugar fermentation by these clostridia grown with combined nitrogen. Hence, molybdenum is probably essential for the nitrogen-fixation process, and it should be included in nitrogen-free media.

Calcium and manganese

Calcium carbonate was originally included in nitrogen-free media to neutralize the large amounts of organic acids formed during the fermentation. But, when the medium is buffered in other ways, such as by phosphate, it is still necessary to include some calcium (e.g. as $CaCl_2$) for essential metabolic processes (Willis, 1934). Manganese, as well as calcium, plays an important role in the formation and germination of bacterial endospores (Gould and Hurst, 1969) so a salt of this metal should also be incorporated in the media.

Other trace elements

Augier (1956) showed that the addition of a mixture of trace elements improved growth of azotobacter in Winogradsky's standard liquid medium with 1% of glucose. Subsequently, Augier (1957) also used the same trace element mixture for nitrogen-fixing clostridia on the assumption that both types of organism have similar requirements for trace elements. A later version of this trace element mixture is given by Pochon and Tardieux (1962). This consists of (g): potassium molybdate, 0·05; sodium borate, 0·05; cobalt nitrate, 0·05; cadmium sulphate, 0·05; copper sulphate, 0·05; zinc sulphate, 0·05; manganese sulphate, 0·05; ferric chloride, a trace; distilled water, 1 litre. This stock solution should be saturated with CO_2. It is used at a rate of 1 ml/litre of medium. Though it has not been proved that nitrogen-fixing clostridia need trace elements other than those specified in the preceding sections above, it is reasonable to include other trace elements for it cannot be assumed that all the nitrogen-fixing anaerobes have the same requirements for them. Elements known to be essential such as molybdenum, can be conveniently incorporated in trace element mixtures, as in that of Pochon and Tardieux (1962) described above.

Organic constituents of nitrogen-free media

Carbon source

Glucose has usually been employed in these media but some strains of clostridia are inhibited by even slight caramelization of this sugar. Sucrose, which is readily utilized by all strains tested in this laboratory, does not caramelize and is therefore safer to use.

Growth factors

All clostridia need organic growth factors for growth (Wilson and Miles, 1957) and sporulation (Perkins, 1965). However, the synthetic nitrogen-free media of classical type do not contain such compounds, so the nitrogen-fixing anaerobes may grow in them only when sufficient growth factors are carried over with the inoculum.

Biotin and yeast extract

Biotin alone was needed for growth of all *Cl. butylicum* strains tested in medium with combined nitrogen, but most *Cl. acetobutylicum* strains tested needed both biotin and *p*-amino-benzoic acid. Yeast extract at 2×10^{-3} μg/ml of medium provided the necessary factors for growth of these *Cl. acetobutylicum* strains (Lampen and Peterson, 1943). Some strains of both these species can fix gaseous nitrogen (McCoy, Higby and Fred, 1928; Rosenblum and Wilson, 1949). According to Carnahan and Castle (1958), *Cl. pasteurianum* needed more biotin when fixing gaseous nitrogen than when growing on sources of combined nitrogen. Other references to growth factors for clostridia are given by Parker (1954).

Yeast extract, which is widely used as a source of growth factors, has rarely been employed in synthetic nitrogen-free media. Hart (1955) used yeast extract, which at 0·02% improved the growth from large inocula and allowed growth from small inocula. Yeast extract was also used by Emtsev (1962).

Soil extract

Augier (1956) found that soil extract at 10 ml/litre of medium improved his medium for azotobacter, and later he included it in his medium for nitrogen-fixing clostridia (Augier, 1957). This extract was made by heating equal weights of garden soil (pH of c. 7·0) and water at 130° for 1 h. It was then filtered through paper, bottled, autoclaved and stored.

Potato extract

Jensen and Spencer (1947) found that small inocula of strains of *Cl. butyricum* grew rapidly and vigorously when the medium was supplemented either with potato extract or with a potato extract concentrate from which most of the nitrogenous and mineral constituents had been removed. This concentrated extract was prepared as follows. Five hundred g of peeled, finely chopped potatoes were steamed for 3–4 h with 1 litre of tap water. The liquid was strained off and the residue washed with hot water to make 1 litre of extract which was filtered with a Buchner funnel. The filtrate was evaporated on a water bath to c. 50 ml, 600 ml of ethanol were added, allowed to stand for at least 24 h and the precipitate removed by filtration. The alcohol was distilled off, the extract brought to 50 ml, autoclaved and stored in the cold. The concentrate was used at 0·2 to 0·5% of the mineral salts—glucose medium (Table 1) which also contained 1 μg each of biotin and *p*-aminobenzoic acid/litre.

Fresh sterile potato sap also initiated growth of *Cl. butyricum* strains when it was added in small amounts to nitrogen-free agar (Parker, 1954).

The use of liquid media for counting nitrogen-fixing clostridia, and for isolation without prior enrichment

Though modified Winogradsky solution was used to detect these organisms in soils (Jensen, 1940) and potato medium was used to count them (Jensen, 1951; Meiklejohn, 1956), the use of synthetic liquid media in Most Probable Number (MPN) counting methods is comparatively recent. Augier (1957) used synthetic liquid medium in test tubes containing Durham tubes to detect gas production. Such tubes are inoculated with different serial dilutions of soil, with enough replicates at each dilution to satisfy the requirements of the chosen method of MPN estimation, and incubated anaerobically at 30°. The following method, modified from Augier according to experience in this laboratory, is recommended for MPN estimation and for isolation. Soil dilutions of 10^{-3}, 10^{-4}, and 10^{-5}, used at 1 ml/tube, give satisfactory results with most soils.

Liquid medium

This medium is basically that of Burk (1930) supplemented with trace elements and growth factors in accordance with the investigations recorded above. The medium comprises: K_2HPO_4, 0·8 g; KH_2PO_4, 0·2 g; $MgSO_4$. $7H_2O$, 0·2 g; NaCl, 0·2 g; $FeSO_4.7H_2O$, 0·01 g; $MnSO_4.7H_2O$, 0·01 g; $CaCl_2$, 0·01 g; glucose (or sucrose), 10·0 g; yeast extract (e.g. Difco), 0·001–0·01 g; $Na_2MoO4.2H_2O$, 0·025 mg; trace element mixture, 1 ml; soil extract, 10 ml; sodium thioglycollate, 1·0 g; distilled water to 1 litre.

When this mixture of phosphates is used the pH will be buffered at c. 7·2. The trace element mixture of Pochon and Tardieux (1962), already described, can be used: the small amounts of iron and molybdenum in it may be omitted. The concentration of thioglycollate given (1 g/litre) increases the number of positive tubes made from high soil dilutions.

Culture vessels

Ten ml portions of the liquid medium are transferred to narrow test-tubes (15 × 150 mm) fitted with Durham tubes, capped or plugged, and autoclaved at 121° for 15 min. Unless the tubes are to be used at once, they should be steamed to remove dissolved air and cooled before inoculation. Inoculated tubes are incubated at 28–30° in nitrogen (see below) for up to 30 days. Results are commonly recorded for counting purposes at 15 and 30 days but positive tubes from which isolations can be made may occur in less than 7 days.

Positive tubes are indicated by abundant gas. Augier (1957) recommends that there should be at least 5 mm of gas in the Durham tube: smaller amounts of gas can be formed by the fermentation of sugar permitted by trace amounts of combined nitrogen in the medium. The medium becomes turbid, with or without a viscid whitish deposit or surface pellicle, and develops an odour of butyric acid or butanol or both.

Isolation

Small inocula from positive tubes made from very dilute suspensions of soil are streaked on plates of nitrogen-free agar medium, which are then incubated anaerobically in an atmosphere of nitrogen. Colonies can usually be picked off within 7 days at 28–30° and purified by repeated streaking on nitrogen-free agar. Inocula can be pasteurized at any stage provided spores are present.

Nitrogen-free agar medium

The liquid medium given above is solidified with 1·5% of agar. Although commercial agars contain combined nitrogen (Table 2), the amounts intro-

TABLE 2. Total nitrogen content of different agar samples

Type of agar	Nitrogen content (%)*
Oxoid Agar No. 1	0·060
Oxoid Ionagar No. 2	0·072
Difco Noble Agar	0·077
Difco Bacto Agar	0·098

* Nitrogen expressed as % of agar taken directly from the bottle.

duced into media are too small to destroy their selectivity for nitrogen-fixing organisms. For example, Bacto Agar with 0·098% of combined nitrogen, and used at 1·5%, provides only 14·7 mg of combined nitrogen/litre of medium. It should be noted that agars contain growth factors and probably inhibitors, of different kinds.

Direct isolation by plating or streaking on nitrogen-free agar medium

The agar medium specified can be used for direct isolation from soil suspensions but, because of the small proportion of nitrogen-fixing anaerobes to other soil organisms, more stages of purification by streak- or pour-plate may be needed than when purifying isolates from enrichment cultures or positive MPN tubes. Estimates of the numbers of nitrogen-fixing clostridia in soils by direct plating on nitrogen-free agar are often small (Swaby, 1939). This may often reflect the true state of affairs in soils, but small counts may also be caused by inability of many cells to grow on the medium. Thus, Parker (1954), using strains of *Cl. butyricum*, found that not all viable cells, as determined by motility, were able to grow on agar media, but most of them did so when fresh potato sap was added to the media. He suggested that the potato sap acted by neutralizing a growth inhibitor present in the agar.

Maintenance of cultures

Jensen and Spencer (1947) used potato medium for this purpose. Test-tubes received 12–15 ml of tap water, *c.* 5 cm deep layer of small pieces of peeled potato, and a little chalk, and were sterilized by autoclaving. If not used at once, the tubes should be heated to boiling point and cooled shortly before inoculation. Such tubes support good growth of nitrogen-fixing clostridia, or non-nitrogen-fixing butyric acid—butanol clostridia, even when incubated in air. Cultures of *Cl. butyricum* usually show strong fermentation at 24 h at 30° or 35°. This potato medium is not suitable for growing pure cultures of *Cl. pasteurianum* which does not ferment starch (see McCoy *et al.*, 1930).

Sometimes a little soil was added by Jensen and Spencer (1947) to the tubes before sterilization to provide an improved maintenance medium.

More convenient maintenance media are RCM (Oxoid) or cooked meat medium in small screw-capped bottles.

Anaerobiosis with steel wool

Alkaline pyrogallol has often been used to remove oxygen from anaerobic incubation vessels to leave an atmosphere of nitrogen favourable for the nitrogen-fixing anaerobes. Unfortunately, hydroxides absorb carbon dioxide which is necessary to initiate growth of some clostridia (Valley and Rettger, 1927; Parker, 1954). Absorption of CO_2 can be prevented by using carbonates or bicarbonates with the pyrogallol instead of hydroxides (Rockwell, 1924) but these mixtures do not absorb oxygen rapidly (Nicol, 1929). Furthermore, all alkaline solutions of pyrogallol of strengths likely to be effective in removing oxygen from air evolve carbon monoxide when oxygen is absorbed (Nicol, 1929), and because this inhibits nitrogen-fixation, at least in aerobes, it is probably better not to use alkaline pyrogallol when growing anaerobic nitrogen-fixers. Hydrogen is not suitable as an anaerobic atmosphere for these clostridia because it inhibits nitrogen fixation (Burris and Wilson, 1945).

Parker (1955) described a method that overcame these difficulties by using activated steel wool to remove oxygen from anaerobic culture chambers. The method depends on the formation of iron-copper couples by the irregular deposition of numerous spots of metallic copper on the surface of the iron. Such activated iron absorbs oxygen readily, though the rate of absorption decreases with decreasing partial pressure of oxygen. Parker also advocated the use of a carbon dioxide supply within the anaerobic culture vessel.

Method

A loose pad of *c*. 10 g of commercial steel wool (grade O or 1) is dipped in 500 ml of the copper sulphate activating solution for a few seconds, drained, placed in a Petri dish lid or open polythene bag, and transferred to the anaerobic jar. Steel wool should be immersed in the copper sulphate solution just long enough to become dark grey and with no more than a trace of copper colour. Wool so treated is irregularly plated with copper, and therefore plentifully supplied with iron-copper couples. Uniformly copper-plated catalysts should not be used.

Commercial steel wool is always greasy, some samples so greasy that the activating solution does not make contact with the metal despite its low surface tension, and the pads do not change colour. These samples of steel wool need at least partial degreasing by washing with carbon tetrachloride, draining, and drying in air. Degreased pads need only a very brief immersion in the copper sulphate solution. Each batch of steel wool should be tested and the correct activating treatment judged.

The activating solution (Parker, 1955)

This comprises: $CuSO_4.5H_2O$, 2·5 g; Tween 80 (or Lissapol), 2·5 g; distilled water, 1 litre: acidified with NH_2SO_4 to pH 1·5–2·0. Cheaper detergents can be used but those specified were selected because of their known lack of toxicity to microorganisms (C. A. Parker, personal communication). Because of the large amount of solution required, and its tendency to give a precipitate on storage, the reagents are best stored as concentrated (e.g. 10%) solutions and mixed when required. The activating solution would then be made by mixing 10% (w/v) solution of $CuSO_4$.$5H_2O$, 25 ml; 10% (w/v) solution of Tween 80, 25 ml; $4NH_2SO_4$, 15 ml; distilled water to 1 litre.

Source of carbon dioxide

Equal weights of magnesium carbonate and sodium bicarbonate are mixed and stored dry. About 5 g of this mixture is placed with 10–15 ml of water in a small beaker and stood in the (3 litre) anaerobic jar. The proportion of water to dry mixture is not important but some solid must remain. Carbon dioxide is slowly released during incubation.

Indicator of anaerobiosis

A mildly alkaline indicator is prepared by mixing two solutions: Solution A comprises 0·5% aqueous methylene blue solution, 3 ml; distilled water to 100 ml: Solution B comprises glucose, 0·5 g; thymol, a small crystal; distilled water to 100 ml; pH adjusted to 10 with 0·1 M carbonate-bicarbonate buffer. The buffer consists of Na_2CO_3 (anhydrous), 1·06 g; $NaHCO_3$, 0·84 g; distilled water to 100 ml. Solutions A and B should be stored in the dark.

Solutions A and B are mixed to give a light to medium blue solution. This may be used cold but will remain coloured until the partial pressure of oxygen inside the anaerobic jar has fallen below 0·05 atm. Many hours usually elapse before cold indicator decolorizes and during this time the operator remains uncertain whether anaerobic conditions are developing.

A more satisfactory method is to boil the indicator to decolorize it, as with McIntosh & Fildes indicator, before placing in the jar. The jar is then filled with nitrogen, and the last traces of oxygen are removed by the activated steel wool. A transient blue colour may appear before the final decolorization.

Cellulose-decomposing Clostridia

These organisms can be isolated by inoculating selective media with dilutions of a soil suspension, and then making subcultures from those liquid cultures or colonies that show digestion of cellulose. Because of the small numbers of cellulolytic anaerobes in many soils, isolation is often easier when preceded by enrichment. However, as already noted for the nitrogen-fixing clostridia, enrichment tends to favour faster growing strains, a disadvantage when knowledge of the range of types present is needed (see Skinner, 1965).

Enrichment cultures

Medium

$(NH_4)_2SO_4$, 1·0 g; K_2HPO_4, 1·0 g; $MgSO_4.7H_2O$, 0·5 g; $CaCO_3$, 2·0 g; NaCl, trace; distilled water, 1 litre (Omelianski, 1902).

Culture vessels

Glass-stoppered reagent bottles (100 ml) each with *c.* 2 g of chopped filter paper.

Procedure

Each bottle is filled nearly full with culture solution, plugged with cotton wool and autoclaved at 121° for 15 min (final pH value of *c.* 7·1). When cold, each bottle is inoculated with *c.* 1 g of soil, enough sterile culture solution added to fill the bottle completely, the sterile glass stopper lightly fitted, the whole inverted momentarily to mix the contents, and incubated at 28–35°.

The primary enrichment culture will contain many organisms able to grow on substrates other than cellulose in the soil inoculum. These organisms will quickly remove remaining traces of oxygen from the medium and make it sufficiently reducing for the cellulolytic anaerobes to grow.

As soon as digestion of the filter paper begins (usually 7 days), as indicated by the appearance of clear patches on the paper, a piece of the paper is transferred to a screw-capped bottle containing anaerobic diluent and

G

nitrogen gas phase (see below), and shaken to disperse the material. The supernatant fluid is used as inoculum for further cultures.

A fresh enrichment culture bottle inoculated with a piece of partly digested paper is likely to show no growth because the medium lacks growth factors and conditions are not sufficiently reducing. For the same reason, too small an inoculum for the primary enrichment culture is undesirable.

Isolation from soil or enrichment culture

A series of culture vessels, containing selective medium with cellulose as sole carbon source, is inoculated with dilutions of soil suspension or primary enrichment culture, and incubated under strict anaerobic conditions. Because very small inocula are used, the medium must contain growth factors and reducing agents. This method of isolation can be modified to give a MPN estimate of numbers of cells present in the inoculum. Liquid or solid media can be used. Inocula can be pasteurized at any stage in the purification procedure.

The media and methods to be described are specific for cellulolytic anaerobes rather than for cellulolytic clostridia *per se*. However, cellulolytic anaerobes in soils are usually spore-formers whereas non-spore-forming types are more characteristic of the rumina of herbivores (see Hungate, 1950).

Liquid medium

Liquid cellulose medium B (Skinner, 1960) is suitable. This consists of buffered mineral salts solution, 500 ml; cellulose suspension (2% w/v), 200 ml; resazurin solution (0·1% w/v), 1 ml; yeast extract (Difco), 1 g; cysteine hydrochloride, 0·5 g; distilled water to 1 litre.

The buffered mineral salts solution comprises: $(NH_4)_2SO_4$, 1 g; $MgSO_4$. $7H_2O$, 0·1 g; NaCl, 2 g; $CaCl_2$, 0·1 g; K_2HPO_4, 13·0 g; KH_2PO_4, 7·0 g; distilled water, 1 litre; pH 7·0. This solution is strongly buffered with phosphate to prevent the cultures becoming very acid, but the phosphate content can be decreased considerably (e.g. to a tenth of the concentrations given above) without affecting attack on the cellulose.

Cellulose suspension prepared by ball-milling (Hungate, 1950) is convenient to use. Growth factors supplied by the yeast extract may be supplemented or replaced by rumen fluid (Hungate, 1950) or other materials (Davies, 1968) when required. Medium without cellulose can be used as anaerobic diluent.

The medium, without the cysteine, is boiled to expel air, cooled to room temperature with a current of oxygen-free nitrogen bubbling through it, the cysteine added, and then distributed to the culture vessels from which the air has been displaced with nitrogen. Suitable apparatus for preparing the medium and dispensing it to the culture vessels has been described in detail (Skinner, 1960). Other similar medium distribution apparatus has been reported by Moore and Cato (1963).

Culture vessels

Screw-capped bottles which can be sterilized with the caps tight after filling with deoxygenated and reduced medium, are more convenient to use than test tubes with rubber stoppers that must be filled with previously sterilized medium.

The black rubber cap liners usually supplied with the bottles are satisfactory, but liners of butyl rubber, which were recommended for anaerobic tube stoppers by Hungate, Smith and Clarke (1966), are available (William Freeman & Co. Ltd., Suba-Seal Works, Staincross, Barnsley, England). Screw caps should be tightened, and may easily be removed, with a tool made from a large rubber stopper (Mackie and McCartney, 1956) or, preferably, a rubber Screw Cap Remover (Johnsons of Hendon Ltd.) (Fig. 4).

After inoculation, each bottle is flushed with sterile oxygen-free nitrogen (c. 750 ml/min) through a bent Pasteur pipette (Hungate, 1950), and using the loose cap to prevent easy access of air (Fig. 1). The cap is then quickly re-fitted and the bottle incubated. When digestion of the cellulose has started, sub-cultures can be made to fresh medium.

Liquid medium can also be used in test-tubes. The medium is prepared in a flask, without being deoxygenated, and 10 ml portions transferred to narrow test-tubes (15 × 150 mm). The tubes are then autoclaved, cooled and inoculated. After inoculation, the tubes are incubated in hydrogen in a McIntosh & Fildes jar in the usual way. Glass jars are more useful than metal jars as they permit observation of the cultures during the rather lengthy incubation period, 2–4 weeks. After sealing the jar, the medium becomes rapidly reduced as indicated by decolorization of the resazurin in the medium. These tubes can be provided with cellulose as particles, as filter paper in pieces, or in long strips (Omelianski, 1902; Fred and Waksman, 1928): air entrained in the paper escapes from the open tubes during autoclaving.

Solid medium

A suitable medium is prepared by using liquid cellulose medium B solidified with 1·5% of agar, and including a little sodium carboxymethyl cellulose to keep the cellulose particles suspended while the agar is setting (Skinner, 1960).

Solid medium can be used with screw capped bottles as follows:

(a) The molten agar (5 ml/28 ml bottle) is allowed to set with the bottle horizontal to form a shallow strip of agar medium along the side. Inoculum can be incorporated in the molten agar or streaked on the surface after solidification (Figs 2 and 4).

FIG. 1. Use of bent Pasteur pipette for flushing screw-capped bottles with nitrogen before caps are tightened. The apparatus is held by the operator during this operation but, for clarity, hands are omitted in the photograph.

(b) The molten medium, with the agar content raised to 2% (5 ml/28 ml bottle), is inoculated and the bottle then rotated until the medium is set in a thin layer on the wall. This roll-bottle culture is then incubated. The "Rolamix" Model RS/54 Roll Tube Culture Machine with a roller speed of 90–100 rev/min (Luckham Ltd., Labro Works, Victoria Gardens, Burgess Hill, Sussex) is suitable for this purpose.

(c) Inoculated molten agar (10 ml/100 ml flat-sided bottle) is allowed

FIG. 2. Screw-capped bottles (28 ml) each with 5 ml of cellulose agar set as a layer parallel to the side of the bottle. Left, generalized digestion of cellulose caused by diffuse growth from soil inoculum on and within the medium. Right, some generalized cellulose digestion, and (top right), localized zone of digestion caused by a single discrete colony (arrowed).

to set as a shallow layer on one of the larger sides and then incubated horizontally with the agar side uppermost (Fig. 3).

All these cultures have a gas phase of oxygen-free nitrogen provided as explained above.

In these cultures, cellulolytic anaerobes develop either as punctiform colonies or as thin-spreading colonies on the surface or at the glass-agar interface. There are two disadvantages in the use of individually sealed

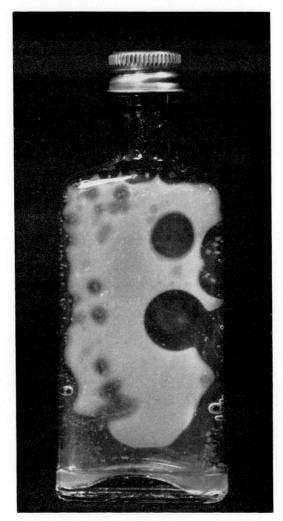

FIG. 3. Flat-sided screw-capped bottle (100 ml) with 10 ml of cellulose agar. Localized digestion of cellulose by small colonies on and within agar. Also, diffuse growth at glass-agar interface.

vessels of solid medium, at least when dealing with cellulolytic anaerobes from soil. Growth is slower in agar, than in liquid, medium, and the organisms tend to spread easily over the agar so it is sometimes difficult to isolate a pure culture from a single colony (Figs 2 and 4).

Cellulolytic anaerobes can be isolated conveniently by an alternative method of streaking inoculum on solid medium in Petri dishes. Medium is

Fig. 4. Left, generalized digestion of cellulose but no colonies visible. This effect is probably caused by the accumulation of cellulolytic enzymes in the condensation water, and often occurs when bottles are incubated vertically. Bottles should be incubated horizontally with the agar uppermost. Right, rubber screw cap remover.

prepared and bottled for storage in 15–20 ml portions in 28 ml bottles. When required, the medium is melted, the contents of one bottle transferred to each dish, and allowed to set. Pour-plates or streak-plates can be used. When set, plates are removed to a McIntosh and Fildes jar which is then filled with hydrogen ($+ c.$ 10% of CO_2) in the usual way. The medium, which becomes pink in colour while the plates are setting on the bench, rapidly decolorizes in the hydrogen atmosphere. Good growth often occurs in $c.$ 7 days at $35°$ on such plates, and there is less tendency for the organisms to spread than in the sealed bottles. A streak plate, 10 days old, is shown in Fig. 5.

An anaerobic jar provides the same atmosphere for all the cultures within it, an advantage when varying other factors likely to affect growth of the organisms being studied. A disadvantage of the jar technique, for liquid and solid medium cultures, is that one culture cannot be removed for study without making it and all the remaining cultures aerobic, at

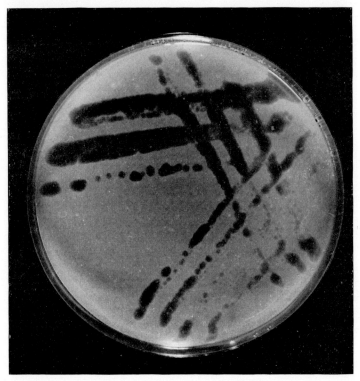

FIG. 5. Streak plate of cellulolytic clostridium culture on cellulose agar. Ten-day-old culture, incubated at 35° in hydrogen in a McIntosh and Fildes jar.

least for a while. This disadvantage can be overcome to some extent by the following procedures. As soon as a culture has been removed from the jar, replace the lid without fixing the clamp, and pass a current of oxygen-free nitrogen through the jar to displace any air that has entered. One litre of nitrogen/min is an adequate flow rate. Alternatively, have ready on the bench an empty jar with ungreased lid in place and pass a current of sterile nitrogen through it to make an oxygen-free reservoir. When the culture jar is opened, the required culture is removed and the remaining cultures placed at once in the nitrogen reservoir. This reservoir jar can easily be converted into the new culture jar should continued incubation be necessary.

Final purification of isolates

Purification of strains can be tedious because of slow growth on cellulose agar. However, growth is more rapid in agar medium containing c. 1% of cellobiose as carbon source in place of cellulose particles, but, because it is less selective than cellulose medium it should be used only in the later stages of purification.

Maintenance of cultures

Pure cultures obtained by different combinations of the methods discussed above can be maintained in liquid cellulose medium. Such cultures in screw-capped bottles remain viable for several months when stored at room temperature.

Supplies of Nitrogen

Oxygen-free nitrogen is readily obtainable in the U.K. but, if only a less pure product is available then it should be passed over heated copper and then through acidified chromous sulphate solution to free it from residual oxygen as recommended by Hungate (1950). It is always advisable to wash nominally oxygen-free nitrogen with acidified chromous sulphate solution, not only to remove traces of oxygen but also any ammonia that can invalidate work with nitrogen-fixing organisms (see Millbank, 1969).

The preparation of chromous sulphate in a container that permits satisfactory sparging of the nitrogen and also rapid reduction of any chromic sulphate formed, is not easy. Gas washing bottles of the Drechsel type can be used but experience in this laboratory has shown that such systems easily leak to the atmosphere with consequent oxidation of the reagent and inability to reduce it. The dual container apparatus described below allows reliable production and storage of chromous sulphate solution. This device has operated well for several years and can be recommended.

Construction of apparatus

The apparatus consists of a gas washing bottle (A) of c. 250 ml capacity fitted with a rubber stopper carrying a vertical glass tube (B) which supports an inverted 350 ml conical flask (C). A Suba-Seal lightly coated with

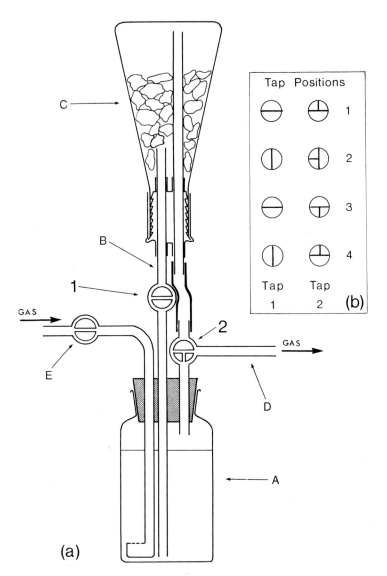

FIG. 6. (a) Apparatus for washing nitrogen to free it from traces of oxygen. A, gas washing bottle (*c.* 250 ml) containing acidified chromous sulphate solution; B, rigid glass tube linking the two containers; C, 350 ml conical flask containing amalgamated granulated zinc; D, gas outlet tube; E, stopcock on inlet side of apparatus. (b, inset) Positions of the stopcocks 1 and 2 depicted in Fig. 6 (a). Diagrams show positions of channels in the stopcock keys. Tap Position (TP) 1, storage position, chromous sulphate solution in reservoir (C); TP 2, gas washing bottle (A) being filled with chromous sulphate solution from reservoir (C), and nitrogen from (A) passing up to (C); TP 3, incoming nitrogen being washed in (A); TP 4, reservoir (C) being filled with solution from (A) prior to storage.

silicone grease makes a good gas-tight seal for (C). The two vessels are inter-connected as shown in Fig. 6. The whole apparatus is supported in a stand.

Charging the apparatus

The reservoir (C), which contains pieces of amalgamated granulated zinc (see Hungate, 1950), is flushed with nitrogen (Tap Position (TP) 4; see Fig. 6b). Then, with TP 3, an acidified solution of potassium chromium sulphate (c. 15 %) is placed in the bottle (A) and nitrogen sparged through it to displace dissolved air. The solution is then forced up into the reservoir (C) using TP 4. The nitrogen flow is then stopped and the apparatus left (TP 1) until the chrome alum is reduced to chromous sulphate, as in-dicated by the development of a brilliant blue colour (c. 2 days). With this tap arrangement, hydrogen (from the action of acid on the zinc) can escape through the outlet tube (D) and a subsequent delivery tube dipping just beneath the surface of mercury. Entry of air back into the apparatus is thereby prevented. One charge of chromous sulphate and zinc should last at least 3 months when used to wash nominally oxygen-free nitrogen.

Using the apparatus

Flush the bottle (A) with nitrogen (TP 3), interrupt the gas flow, and then fill bottle (A) by using TP 2, which allows the solution to pass down into (A) while the displaced nitrogen is forced back into the reservoir (C). When (A) is filled to a suitable level, pass the nitrogen to be washed through it (TP 3) as long as required. When gas washing is finished, force the solution up into the reservoir (TP 4) until it is full, change to TP 1, and stop the gas flow. Any chromic ions formed in (A) while gas is being washed are rapidly reduced to the chromous state when the solution is brought into contact again with the large surface of zinc in (C). During gas washing, the outlet tube (D) is connected to a Drechsel bottle containing distilled water to trap spray from the bottle (A). The inlet tap (E) is closed when the apparatus is not in use. This tap is conveniently preceded by a flow meter, and a needle control valve which can be opened quickly to vent the gas supply to atmos-phere whenever tap manipulations prevent free flow of gas.

Acknowledgements

I am indebted to Dr C. A. Parker for helpful discussion concerning the preparation of activated steel wool, and to Mr F. G. Hamlyn for performing the nitrogen analyses of agars given in Table 2. Also, I thank Dr H. L. Jensen and the Linnean Society of New South Wales for permission to reproduce a short extract from the paper by Jensen and Spencer (1947).

References

AUGIER, J. (1956). A propos de la numération des *Azotobacter* en milieu liquide. *Annls Inst. Pasteur, Paris,* **91,** 759.

AUGIER, J. (1957). A propos de la fixation biologique de l'azote atmosphérique et de la numération des *Clostridium* fixateurs dans les sols. *Annls Inst. Pasteur, Paris,* **92,** 817.

BREED, R. S., MURRAY, E. G. D., & SMITH, N. R. (1957). *Bergey's Manual of Determinative Bacteriology,* 7th ed. London: Baillière, Tindall & Cox Ltd.

BURK, D. (1930). The influence of nitrogen gas upon the organic catalysis of nitrogen fixation by Azotobacter. *J. phys. Chem.,* **34,** 1174.

BURRIS, R. H. & WILSON, P. W. (1945). Biological nitrogen fixation. *A. Rev. Biochem.,* **14,** 685.

CARNAHAN, J. E. & CASTLE, J. E. (1958). Some requirements of biological nitrogen fixation. *J. Bact.,* **75,** 121.

DAVIES, M. E. (1968). Role of colon liquor in the cultivation of cellulolytic bacteria from the large intestine of the horse. *J. appl. Bact.,* **31,** 286.

EMTSEV, V. T. (1962). Quantitative estimation of anaerobic nitrogen-fixing butyric acid bacteria belonging to the genus *Clostridium* in soil. *Mikrobiologiya,* **31,** 288.

FRED, E. B. & WAKSMAN, S. A. (1928). *Laboratory Manual of General Microbiology.* New York: McGraw-Hill Book Co., Inc.

GIBBS, B. M. & FREAME, B. (1965). Methods for the recovery of clostridia from foods. *J. appl. Bact.,* **28,** 95.

GOULD, G. W. & HURST, A. (1969). *The Bacterial Spore.* London & New York: Academic Press.

HART, M. G. R. (1955). A study of sporeforming bacteria from soil. Ph.D. Thesis, University of London.

HIRSCH, A. & GRINSTED, E. (1954). Methods for the growth and enumeration of anaerobic spore-formers from cheese, with observations on the effect of nisin. *J. Dairy Res.,* **21,** 101.

HUNGATE, R. E. (1950). The anaerobic mesophilic cellulolytic bacteria. *Bact. Rev.,* **14,** 1.

HUNGATE, R. E. SMITH, W. & CLARKE, R. T. J. (1966). Suitability of butyl rubber stoppers for closing anaerobic roll culture tubes. *J. Bact.,* **91,** 908.

JENSEN, H. L. (1940). Contributions to the nitrogen economy of Australian wheat soils, with particular reference to New South Wales. *Proc. Linn. Soc. N.S.W.,* **65,** 1.

JENSEN, H. L. (1951). Notes on the microbiology of soil from Northern Greenland. *Meddr Grønland,* **142,** 23.

JENSEN, H. L. & SPENCER, D. (1947). The influence of molybdenum and vanadium on nitrogen fixation by *Clostridium butyricum* and related organisms. *Proc. Linn. Soc. N.S.W.,* **72,** 73.

LAMPEN, J. O. & PETERSON, W. H. (1943). Growth factor requirements of clostridia. *Archs Biochem.*, **2**, 443.

MACKIE, T. J. & McCARTNEY, J. E. (1956). *Handbook of Practical Bacteriology*, 9th ed. Edinburgh & London: E. & S. Livingstone Ltd.

McCOY, E., HIGBY, W. M., & FRED, E. B. (1928). The assimilation of nitrogen by pure cultures of *Clostridium pasteurianum* and related organisms. *Zentbl. Bakt. ParasitKde*, Abt. II, **76**, 314.

McCOY, E., FRED, E. B., PETERSON, W. H. & HASTINGS, E. G. (1930). A cultural study of certain anaerobic butyric-acid-forming bacteria. *J. infect. Dis.*, **46**, 118.

MEIKLEJOHN, J. (1956). Preliminary notes on numbers of nitrogen fixers on Broadbalk field. *Proc. 6th Int. Congr. Soil Sci., Paris.*, **C**, 243.

MILLBANK, J. W. (1969). Nitrogen fixation in moulds and yeasts—a reappraisal. *Arch. Mikrobiol.*, **68**, 32.

MOORE, W. E. C. & CATO, E. P. (1963). Validity of *Propionibacterium acnes* (Gilchrist) Douglas and Gunter Comb. Nov. *J. Bact.*, **85**, 870.

NICOL, H. (1929). Note on anaerobiosis and the use of alkaline solutions of pyrogallol. *Biochem. J.*, **23**, 324.

OMELIANSKI, W. (1902). Über die Gärung der Cellulose. *Zentbl. Bakt. ParasitKde*, Abt II, **8**, 225.

PARKER, C. A. (1954). Non-symbiotic nitrogen-fixing bacteria in soil. I. Studies on *Clostridium butyricum*. *Aust. J. agric. Res.*, **5**, 90.

PARKER, C. A. (1955). Anaerobiosis with iron wool. *Aust. J. exp. Biol. med. Sci.*, **33**, 33.

PERKINS, W. E. (1965). Production of clostridial spores. *J. appl. Bact.*, **28**, 1.

POCHON, J. & TARDIEUX, P. (1962). *Techniques d'Analyse en Microbiologie du Sol.* St. Mandé (Seine): Éditions de la Tourelle.

ROCKWELL, G. E. (1924). An improved method for anaerobic cultures. *J. infect. Dis.*, **35**, 580.

ROSENBLUM, E. D. & WILSON, P. W. (1949). Fixation of isotopic nitrogen by clostridium. *J. Bact.*, **57**, 413.

SKINNER, F. A. (1960). The isolation of anaerobic cellulose-decomposing bacteria from soil. *J. gen. Microbiol.*, **22**, 539.

SKINNER, F. A. (1965). The enrichment and isolation of anaerobic cellulolytic soil bacteria. In *Anreicherungskultur und Mutantenauslese*, (Symp. Deutsch. Gesellsch. Hyg. Microbiol.). *Zentbl. Bakt. ParasitKde*, Abt I., Suppl. Heft 1, 91.

SWABY, R. J. (1939). The occurrence and activities of *Azotobacter* and *Clostridium butyricum* in Victorian soils. *Aust. J. exp. Biol. med. Sci.*, **17**, 401.

VALLEY, G. & RETTGER, L. F. (1927). The influence of carbon dioxide on bacteria. *J. Bact.*, **14**, 101.

WAKSMAN, S. A. (1927). *Principles of Soil Microbiology*. London: Baillière, Tindall & Cox.

WILLIS, A. T. (1964). *Anaerobic Bacteriology in Clinical Medicine*, 2nd ed. London: Butterworths.

WILLIS, W. H. (1934). The metabolism of some nitrogen-fixing clostridia. *Res. Bull. Iowa agric. Exp. Stn.*, 173.

WILSON, G. S. & MILES, A. A. (1957). *Topley and Wilson's Principles of Bacteriology and Immunity*. 4th ed. London: Edward Arnold Ltd.

WINOGRADSKY, S. (1893). Sur l'assimilation de l'azote gazeux de l'atmosphère par les microbes. *C. r. hebd. Séanc. Acad. Sci., Paris,* **116,** 1385.

WINOGRADSKY, S. (1895). Rechèrches sur l'assimilation de l'azote libre de l'atmosphère par les microbes. *Archs. Sci. biol., St. Petersb., (Arkh. biol. Nauk.)* **3,** 297.

WINOGRADSKY, S. (1902). *Clostridium Pastorianum,* seine Morphologie und seine Eigenschaften als Buttersäureferment *Zentbl. Bakt. ParasitKde.,* Abt. II, **9,** 43.

Anaerobic Jars in the Quantitative Recovery of Clostridia

B. V. Futter and G. Richardson

*Microbiology Section, School of Pharmacy, Portsmouth Polytechnic,
Portsmouth, England*

The design of anaerobic jars is based on sound principles. However, lack of attention to detail in use has sometimes led to occasional complaints of an inability to reproduce anaerobic conditions. The total lack of information concerning the environment within such jars must in part be responsible for this state of affairs. We embarked on a programme of work in an attempt to describe quantitatively the conditions which develop within an anaerobic jar so that their importance to growth could be assessed.

Firstly an anaerobic jar was modified so that measurements of Eh, pH, pCO_2 and pO_2 could be made before and during growth. Secondly the effect of variation in these conditions on recovery was assessed using colony count by a roll-tube technique (Futter and Richardson, 1970a) as a criterion. This is a more exacting test than those based on other growth measurements which may only show that at least one organism in a population has been able to give rise to a viable clone. Although hydrogen is probably the gas most frequently employed in anaerobic work, nitrogen, methane and North American city gas find some use and it seemed of importance to examine the effect of choice of gas in detail.

The original jar was described by Futter and Richardson (1970b) and further modifications by one of us (GR) have led to the description given here. Additional details on the effect of conditions on recovery will be found in the same two papers and in that of Futter (1967). Except where stated, results have been compiled using control suspensions of *Clostridium welchii* spores and suspensions treated by gamma-rays or heat (0·1% survival) and ethylene oxide (2% survival). In this chapter we summarize the techniques used and discuss the more important findings in the light of our further personal experience.

Use of Anaerobic Jar with Hydrogen

We favour the cold catalyst glass jars (Gallenkamp, Christopher Street, London, E.C.2) because of the distinct advantage of visibility of contents and also freedom from corrosion. These jars have a gas volume of 3·4 litres and comfortably hold 48 roll tubes in two tiers of 24. The element of risk associated with glass jars due to theoretical possibility of explosion is discounted. We have used glass jars extensively for several years without

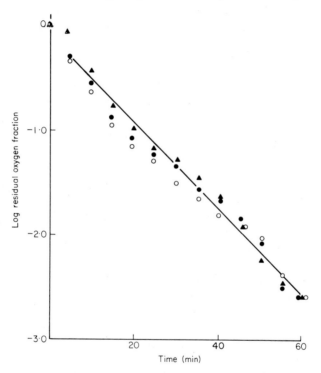

FIG. 1. Rate of combination of hydrogen and oxygen in an anaerobic jar (three replicated experiments).

mishap in routine use. Stokes (1968) adequately describes the use of anaerobic jars but the following points are of interest in precise quantitative work when a high degree of reproducibility of environment is important.

The catalysed reaction between oxygen and hydrogen is slow; even with a new catalyst a one-hundredfold reduction in oxygen concentration takes about 45 min (Fig. 1). Moreover, catalysts are sensitive to poisoning which is particularly likely to occur under the conditions of use (e.g. by evolved gases and condensate) so that the rate of combination may well be much lower. It is neither satisfactory to assess completion of the reaction in terms of a suitably low rate of flow (as seen by bubbling through a wash bottle) of hydrogen nor by Eh measurement. The former may simply indicate that the catalyst is inactive and a more positive check can be easily made on the catalyst itself. The sachet should become too hot to hold when exposed to a jet of hydrogen. In the case of Eh the influence of hydrogen greatly outweighs that of oxygen and a low redox value may be recorded in the presence of residual oxygen irrespective of the state of catalytic activity.

An acceptably low oxygen tension may be achieved in a relatively short time by suitable evacuation procedures, but the continued slow reaction between the gases can result in formation of a vacuum within the jar. This is particularly serious when the jar is used to capacity and contains a large volume of medium. There is then an appreciable reservoir of oxygen dissolved in the medium which is not removed during evacuation and subsequently diffuses back into the gas space. Under such conditions the marked vacuum which develops may result in failure of the seals so that air is drawn back into the jar and failure in anaerobiosis results. Thus we deem it advisable, the more so since it requires no extra effort, to fill with hydrogen via a bladder and to leave the bladder attached to the jar for the first hour or so of incubation. If this procedure is adopted it has the further advantage that a high degree of prior evacuation is not required and a reduction in pressure to 250 mm Hg only has proved satisfactory.

Use of Jar with Nitrogen

For the reasons outlined below it may be preferable to incubate in nitrogen rather than hydrogen. Physical removal of oxygen to an acceptable level (less than 0.1%) is effected by four consecutive evacuations of at least 650 mm Hg and refilling with oxygen-free nitrogen. The gas is stated to be 99.995% pure although from measurements made on the gas before and after passing over heated copper turnings we found one sample of gas to contain a higher percentage of oxygen than indicated by this figure. Occasionally contamination by moulds proved troublesome in nitrogen-filled jars.

Indicators of Anaerobiosis

It is usually advocated that a redox indicator be included in the jar as a check that lack of growth is not due to failure in anaerobiosis. We have on occasion used the Lucas methylene blue indicator (Stokes, 1968) distributed in 5 ml ampoules. These may be stored open in a humidified chamber when

H

not in use and can be used repeatedly for a limited period. Doubt is sometimes expressed regarding the satisfactory nature of the indicator. Correct pH adjustment is necessary in preparation for reproducible behaviour. The indicator at pH 7·8 is 99% reduced and colourless at about −80 mv. However quite different oxygen tensions are required to give the same potential in hydrogen and nitrogen, and decolorization occurs more slowly in the latter. With this in mind it is more satisfactory to include a culture of a known strict anaerobe as indicator each time the jar is set up. For this purpose it is convenient to use a dilute stock suspension of spores of *Clostridium tetani*. The comments of Willis (1964) on biological indicators are not reasonable. If one cannot culture a control organism successfully there is little hope with an unknown.

Measurement of Cultural Conditions

An anaerobic jar was adapted to allow the concurrent measurement of pH and Eh of medium and pO_2 and pCO_2 of gas phase (Fig. 2). The electrodes were connected to a Radiometer (Copenhagen) pH meter Model No. 27 which, with attached gas monitor Model No. 927 enabled direct read-out of all four parameters. The lid of the jar was drilled to take four No. 17 bungs which were cemented in place with Araldite (Ciba, Duxford, Cambridgeshire). Two of these carried the reservoir elements (A) of the reference electrodes (B), the third, the leads of the electrodes (S), and the fourth a three-way tap (J) for connections to gas electrodes.

Reference electrodes

Two reference electrodes were employed so that duplicate readings of redox potential could be obtained or values of pH and Eh simultaneously in separate culture containers. The electrodes were of the silver/silver chloride type with 3·5 M KCl as electrolyte and embodied remote micro-liquid junction tubes (Activion Ltd., Mitchell Hall, Kinglassie, Fife). Electrodes employing saturated KCl are inconvenient in use due to problems associated with crystallization. The electrolyte reservoir mounted on the lid was connected by silicone rubber tubing to the micro-liquid junction component (C). This consisted of a length of 3·5 mm diam glass tubing terminating in a porous ceramic plug. Several such tubes were kept available in a sterile state so that they could be readily interchanged. The rate of diffusion through the plug was stated to be less than 0·005 ml/h so that contamination of medium by KCl was not a problem and there was no appreciable flow through the plug when evacuating the jar during setting-up procedures.

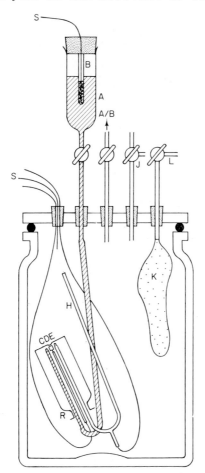

FIG. 2. The anaerobic jar; A, KCl reservoir; B, reference electrode; C, micro-liquid juction tube; D, spade-form platinum electrode; E, spear pH electrode; H, side-arm of pH electrode; J, three-way tap for connection to gas electrodes; K, bladder; L, valve; R, roll-tube containing pH and redox electrodes, and S, electrode leads (one spare for extra Pt electrode). *N.B.* Needle valve for evacuating and filling is not shown.

Redox electrodes

Platinum electrodes in spade form (D) were preferred for redox measurements. The large surface area decreased their susceptibility to poisoning which could become troublesome after several weeks in routine use. Electrodes were cleaned by heating to red heat, plunging in hot *aqua regia* and finally rinsing in distilled water.

pH electrodes

For measurements in roll tubes during growth the electrode must be inverted. With conventional electrodes this usually results in the trapping of a bubble of air against the membrane thus rendering the electrode inoperative. The spear electrode (E) was specially constructed (Activion Ltd.) with a long side arm (H) so that the air bubble could be directed away from the membrane. Additionally the side arm was useful in supporting the electrode diagonally across the jar. The electrode was fitted with the type of cable and plug used on steam sterilizable electrodes. This assists in proper functioning in the damp environment of the jar.

Oxygen electrode

A Clark type electrode mounted in a thermostatted jacket (not shown) was connected to one arm of the three-way tap (J) by a short length of 1·5 mm diam plastic tubing. As the capacity of the electrode cell was only 70 μl very small samples of gas were needed for measurement. These were expelled from the jar as required without affecting the gaseous environment by inflating a small bladder (K) hanging in the jar. (In case of doubt about the permeability of the bladder to oxygen, hydrogen or nitrogen as appropriate could be used to inflate the bladder so that minimum disturbance of gaseous environment is caused.) Care is necessary when setting up the jar, before evacuating, to make sure that the bladder is completely empty and during evacuation that it is isolated by means of a tap. After the gas sample is expelled a valve (L) prevents the bladder from deflating thus sucking air back into the jar via the gas electrode sample tube.

Carbon dioxide electrode

A Severinghaus type electrode (Radiometer) was employed and used as described under "oxygen electrode".

Effect of Environmental Conditions on Viable Count

Eh of medium

Reproducible values for initial Eh are sometimes difficult to obtain; values observed depend considerably on the degree of prior oxidation to which the medium has been subjected. This varies from deep tubes of media to thin films of agar in roll tubes and also depends on the time taken to set up preparations. Nevertheless it is shown in Table 1 and Fig. 3 how the Eh of

TABLE 1. Eh* (mv) of some modifications of RCM agar, 30 min after setting up in roll tubes

Reductant		Gaseous environment		
		H_2	N_2	air
None		−325	+150	+200
Cysteine HC1	0·05%	−340	+155	+200
	0·10%	−340		
	0·15%	−340		
	0·20%	−350		
$Na_2S_2O_4$	0·05%	−355	+140	+170
	0·10%		+140	
	0·15%	−370	+100	
	0·20%	−400	+10	+85
Ascorbic acid	0·10%	−305		
$FeSO_4$	0·05%	−345		

* Final Eh is dependent on period of exposure to air during setting up. Eh of freshly prepared RCM in deep tubes in hydrogen is −410 mv.

media in roll tubes prepared under conditions as similar as possible compares in hydrogen, nitrogen and air. Measurements were made ½ h after inoculation.

Counts of control and treated spore suspensions have been compared over a wide range of Eh (from −410 mv to +155 mv) and found to be similar. This pattern of behaviour is in marked contrast to the variation in count which is gas-dependent (see below) and we conclude therefore that Eh *per se* is not important as a determinant of viable count.

Hydrogen and nitrogen

In all cases investigated the count in nitrogen was at least equal to and often higher than that in hydrogen. Curves illustrating the dependence of recovery on composition of gas mixture are shown in Figs 4 and 5. It is apparent that although normal spores are not fastidious with respect to gas, survivors from exposure to different bactericidal processes can be distinctly exacting in growth requirements. Counts of gamma-irradiated and ethylene-oxide treated spores were respectively threefold and twofold higher in nitrogen than hydrogen.

Although it seems obvious that this is a gas-dependent effect we have not been able to distinguish between the relative importance of nitrogen stimulation and hydrogen toxicity. It seems unlikely that nitrogen fixation is implicated as replacement of nitrogen by argon gave a similar enhanced recovery of irradiated spores when compared with hydrogen.

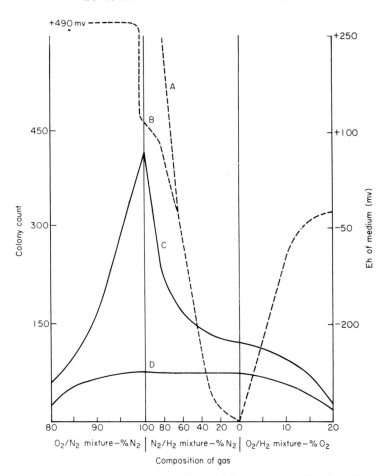

FIG. 3. Variation in the recovery of *Cl. welchii* spores and Eh of medium with gaseous environment. Eh, broken lines; A, commercial nitrogen, and B, oxygen-free nitrogen. Recovery, solid lines; C, survivors from *gamma* irradiation, and D, control spores (counts reduced by 10^{-4})

Heated spores were unusual in that counts in a critical gas mixture of 50% hydrogen and nitrogen were lower than in either gas alone. The argument has been advanced (Futter and Richardson, 1970b) that this represents an isolated case of dependence of count on a region of unfavourable Eh rather than on gas composition.

With undamaged organisms (spores and vegetative cells) *Cl. septicum* counts in nitrogen were up to threefold those in hydrogen. With spore suspensions this effect was most marked immediately after preparation of the suspension and disappeared after about four months' storage at 4°. Irradiated spores of *Cl. septicum, Cl. histolyticum* and *Cl. sporogenes* also showed a "nitrogen dependence" of a similar magnitude.

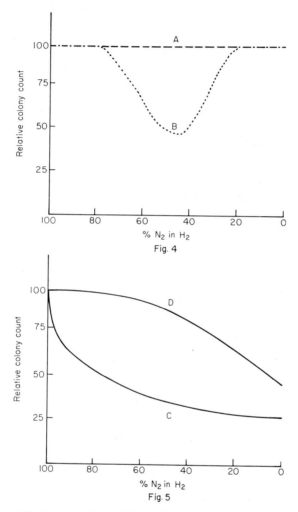

FIG. 4 (upper). Variation of effect of nitrogen on recovery of *Clostridium welchii* with pre-treatment of spores. A, control spore suspension, and, B, heated spores (0·1 % survival).

FIG. 5 (lower). Variation of effect of nitrogen on recovery of *Clostridium welchii* with pre-treatment of spores. C, gamma-irradiated spores (0·1 % survival), and D, ethylene oxide-treated spores (2 % survival).

Oxygen

Cl. welchii is aerotolerant and counts were not affected by up to 5% oxygen (Fig. 3). Further increase in $\% O_2$ in N_2 or H_2 resulted in similar reductions in count in both gas systems although the Eh of the medium in the two cases varied widely. This offers further evidence that Eh is relatively unimportant.

The effect of oxygen on the growth of small inocula of vegetative cells of *Cl. tetani, Cl. septicum* and *Cl. oedematiens* has also briefly been examined. All three behaved similarly in that oxygen tensions of 0·2 to 0·3% were limiting.

Carbon dioxide

Despite the widespread advocacy for the routine inclusion of this gas in anaerobic jars we have found that quite small concentrations (1%) resulted in reduced counts and it seems desirable that quantitative studies should be extended to other organisms.

Conclusions

1. Eh is not a satisfactory basis for the characterization of the anaerobic environment. Eh values are meaningless unless other conditions are exactly specified and Eh itself has little effect on recovery. Definition of the final oxygen concentration and the means whereby it is established, particularly the nature of the replacement gas used, is required in exacting or quantitative work.

2. Recovery in nitrogen is often higher than in hydrogen. With our experience limited to a few species we therefore prefer routine incubation to be in nitrogen rather than in hydrogen. It should be stressed that our criterion is one of maximum quantitative recovery. As more care and time is needed when nitrogen is used, in other contexts this extra effort may not be worthwhile.

3. Even when jars are used with hydrogen care must be exercised in their use and factors likely to result in failure fully appreciated.

Acknowledgement

The assistance of Mr J. H. Shah is gratefully acknowledged in producing the data presented in Fig. 1.

References

FUTTER, B. V. (1967). The detection and viability of anaerobic spores surviving bactericidal influences. Ph.D. thesis (C.N.A.A.) Portsmouth Polytechnic.

FUTTER, B. V. & RICHARDSON, G. (1970a). Viability of clostridial spores and the requirements of damaged organisms. I. Method of count, period and temperature of incubation, pH of medium. *J. appl. Bact.*, **33,** 321.

FUTTER, B. V. & RICHARDSON, G. (1970b). Viability of clostridial spores and the requirements of damaged organisms. II. Gaseous environment and redox potentials. *J. appl. Bact.*, **33,** 331.

STOKES, E. J. (1968). *Clinical Bacteriology. 3rd ed.* London: Arnold.

WILLIS, A. T. (1964). *Anaerobic Bacteriology in Clinical Medicine. 2nd ed.* London: Butterworth.

The Cultivation of Human Intestinal Bacteria

B. S. Drasar and J. S. Crowther

Bacteriology Department, Wright–Fleming Institute, St. Mary's Hospital Medical School, London, England

The bacterial flora of the intestine is very complex. The aerobic bacteria, such as enterobacteria and enterococci, are easy to grow and attention has in the past focused upon them. However, non-sporing anaerobes were first isolated from faeces many years ago (Disato, 1914) and their numerical dominance was demonstrated over thirty years ago (Sanborn, 1931; Eggerth and Gagnon, 1933). Since anaerobic bacteria appear to be of great importance we have tried to develop methods for their quantitative recovery.

Although the aerobic bacteria in faeces probably protect the anaerobic bacteria by the production of an anaerobic micro-environment, preservation of specimens is essential if the relative numbers are to be maintained for any time.

The cultivation of oxygen-sensitive bacteria presents difficulties: using conventional techniques anaerobic bacteria are exposed to oxygen dissolved in diluents and culture media. The spreading of specimens on plates isolates single bacterial cells in the presence of oxygen and thus maximizes its lethal effects. Even in an anaerobic jar diffusion of dissolved gas from agar plates is slow and the exposure of anaerobic bacteria to oxygen is prolonged.

The preparation, dispensing and inoculation of media under a stream of oxygen-free gas, as suggested by Hungate (1950), has considerable advantages, but the manipulatory difficulties of the technique are considerable and, furthermore, plates cannot be used. We have used an anaerobic cabinet to overcome the manipulatory difficulties of the Hungate technique while preventing access of oxygen to the culture media during the preparation of culture plates, inoculation and incubation.

In addition to the preservation and cultivation of specimens the maintenance of pure cultures of non-sporing anaerobes is important.

The Preservation and Transport of Specimens

We have routinely preserved specimens of faeces and intestinal juice frozen as a 1:10 dilution in a glycerol broth containing 1% Lab Lemco (Oxoid) and 10% glycerol (Drasar, Shiner and McLeod, 1969). Detailed bacteriological analysis of replicate samples of faeces taken before freezing and at periods of up to one month after freezing showed no significant change in the viable counts of the various organisms present in the sample (Crowther, 1971, in preparation for *J. appl. Bact.*)

The Anaerobic Cabinet and its use

The cabinet (Fig. 1) consisted of an airtight Perspex glove box fitted with

FIG. 1. The anaerobic cabinet. Within the body of the cabinet can be seen 2 photographic warming plates for drying plates. A converted milking machine pail is in the air-lock. The body of the cabinet (excluding the air-lock) is 3 ft long.

gas-tight doors. The bulk of the oxygen was removed from the cabinet by burning a spirit lamp within it; at the same time the cabinet was flushed at approximately 20 l/min with nitrogen containing 10% carbon dioxide; the final traces of oxygen were removed by recirculating the cabinet's atmosphere through a hot, copper-packed deoxygenating column* (Moore, 1966) at 27 l/min. During culture of bacteria a slight positive pressure was maintained in the cabinet; the gas flow was reduced to 5 l/min.

Stainless steel milking-machine pails with vacuum tight lids (Fullwood Bland and Co.) fitted with vacuum taps and cold catalysts ("D" catalyst, Engelhard Industries) were used in place of conventional anaerobic jars

* See also p. 134.

as suggested by Schaedler, Dubos and Costello (1965). When used in conjunction with the cabinet the pails were evacuated before being placed in the airlock.

Preparation of media for use in the cabinet

Reinforced Clostridial Agar (Oxoid) fortified with 1% liver digest (Oxoid) was used for the culture of the strict anaerobes. The pH was adjusted to 7·5 giving a final pH of 7·3 after autoclaving at 121°/15 min. The medium was prepared in bulk and stored until required. The agar was steamed with the cap loose, the cap being retightened on removal from the steamer. The molten agar was passed into the cabinet, the blood added, and the plates poured; they were dried on a warming plate.

TABLE 1. Media for the isolation of intestinal bacteria

Medium	Incubation Atmos- phere	Days	Organism	Notes, sources and references
Reinforced Clostridial Agar	An	4	Total anaerobes	Oxoid with 1% liver digest and 10% horse blood. pH adjusted to 7·5±0·1 before auto-claving. Plates poured, dried and inoculated in cabinet.
Rogosa's Agar (V)	An	4	*Veillonella*	Rogosa (1956), Rogosa *et al.* (1958)
Tomato Juice Agar	An	1	Anaerobic sarcinae	Oxoid; pH adjusted to 7·0
Willis & Hobbs' Agar	An	4	Clostridia	Willis and Hobbs(1959) Neomycin 40 μg/ml
	An	1	Clostridia, Anaerobic sarcinae	Inoculum heated for 10 min at 70°. No anti-biotics
Azide agar	0	1	Enterococci	Schaedler *et al.* (1965)
Blood agar	1	1	Total aerobes	
MacConkey's Agar	0	1	Enterobacteria, Enterococci	Oxoid
Mannitol Salt Agar	0	3	*Staphyloccus aureus*, *Bacillus* spp	Oxoid, incubate 30°
Rogosa's Agar (L)	90%CO$_2$	3	Lactobacilli	Rogosa *et al.* (1951)
Sabouraud's Agar	0	5	Yeasts, filamentous fungi	Chloramphenicol 40 μg/ml duplicates incu-bated 37° and 22°
S$_1$ Agar	0	2	*Streptococcus salivarius*	Williams and Hirsch (1950)
Nutrient agar	0	1	*Bacillus* spp	Inoculum heated for 10 min at 70°

O, aerobic; An, anaerobic (90% (v/v) H$_2$ and 10% (v/v) CO$_2$). Incubation, 37°.

Brain-heart infusion broth (Oxoid) with L-cysteine HCl (0·05%) was used as a diluent. The broth was steamed before use.

Inoculation of media

Table 1 gives details of the media routinely employed. Specimens were inoculated on to Reinforced Clostridial Agar into the cabinet; the other media were inoculated on the open bench.

Tenfold dilutions of the specimens were prepared in the cabinet. An 0·1 ml sample of each dilution was spread on to the prepared plates; these were then packed into an anaerobic jar in the airlock. The jar was removed and its atmosphere twice replaced by a mixture of 90% hydrogen and 10% carbon dioxide before incubation at 37°.

To count anaerobic sarcinae and spore formers serial dilutions of the specimen were heated for 10 min at 70° before plating out. After anaerobic incubation sarcinae could be readily distinguished (Crowther, 1971).

The Preservation of Strains of Strictly Anaerobic Bacteria

Strains of strictly anaerobic bacteria were preserved by the addition of sterile glycerol to a final concentration of 10% to a 4-day culture of the organism in Robertson's meat broth (Southern Group Labs.). The cultures were stored frozen at $-20°$ or below. Over 80% of the cultures remain viable for at least 9 months. Cooked meat broth prepared from tablets (Oxoid) does not support the growth of non-sporing anaerobes and therefore should not be used.

Discussion

The media and methods described here enable us to count the major broad groups of bacteria present in the intestine. The methods are similar to those described previously (Drasar, 1967) but the use of reheated media in the cabinet simplifies the procedure for the cultivation of oxygen-sensitive bacteria.

Oxygen cannot dissolve in the media or diluents within the cabinet and this reduces the exposure of oxygen sensitive bacteria to its influence. The use of an anaerobic cabinet enables one to isolate numerous oxygen-sensitive bacteria not readily cultivated by conventional techniques (Drasar 1967, Lee, Gordon and Dubos, 1968; Aranki et al., 1969).

References

ARANKI, A., SYED, S. A., KENNEY, E. B. & FRETER, R. (1969). Isolation of anaerobic bacteria from human gingiva and mouse caecum by means of a simplified glove-box procedure. *Appl. Microbiol.*, **17**, 568.

CROWTHER, J. S. (1971). *Sacrina ventrioli* in man. *J. med. Microbiol.* (In press).

DISATO, A. (1914). Contribution à l'étude sur l'intoxication intestinale. *Zentbl. Zentbl. Bakt. ParasitKde,* Abt. I. Orig., **62**, 433.

DRASAR, B. S. (1967). Cultivation of anaerobic intestinal bacteria. *J. Path. Bact.*, **94**, 417.

DRASAR, B. S., SHINER, M. & MCLEOD, G. M. (1969). The bacterial flora of the gastrointestinal tract in healthy and achlorhydric persons. *Gastroenterology*, **56**, 71.

EGGERTH, A. H. & GAGNON, B. H. (1933). The bacteroides of human faeces. *J. Bact.*, **25**, 389.

HUNGATE, R. E. (1950). The anaerobic mesophilic cellulolytic bacteria. *Bact. Rev.*, **14**, 1.

LEE, A., GORDON, J. & DUBOS, R. (1968). Enumeration of oxygen sensitive bacteria usually in the intestine of healthy mice. *Natural Lond.*, **220**, 1137.

MOORE, W. E. C. (1966). Techniques for routine cultivation of fastidious anaerobes. *Int. J. Syst. Bact.*, **16**, 173.

ROGOSA, M. (1956). A selective medium for the isolation and enumeration of the *Veillonella* from the oral cavity. *J. Bact.*, **72**, 533.

ROGOSA, M., MITCHELL, J. A. & WISEMAN, R. F. (1951). A selective medium for the isolation of oral and fecal lactobacilli. *J. Bact.*, **62**, 132.

ROGOSA, M., FITZGERALD, R. J., MACKINTOSH, M. E. & BEAMAN, A. J. (1958). Improved medium for selective isolation of *Veillonella*. *J. Bact.*, **76**, 455.

SANBORN, A. G. (1931). The faecal flora of adults with particular attention to individual differences and their relationship to diet. *J. infect. Dis.*, **48**, 541.

SCHAEDLER, R. W., DUBOS, R. & COSTELLO, R. (1965). The development of the bacterial flora in the gastrointestinal tract of mice. *J. exp. Med.*, **122**, 59.

WILLIAMS, R. E. O. & HIRSCH, A. (1950). The detection of streptococci in air. *J. Hyg., Camb.*, **48**, 505.

WILLIS, A. T. & HOBBS, G. (1959). Some new media for the isolation and identification of clostridia. *J. Path. Bact.*, **77**, 511.

Use of Selective Media for Isolation of Anaerobes from Humans

S. M. FINEGOLD, P. T. SUGIHARA, AND VERA L. SUTTER

Wadsworth General Hospital, Veterans Administration, Los Angeles, California, U.S.A.

Selective media are very useful for facilitating recovery of specific bacteria from complex mixtures of organisms such as are found in faeces, the normal oral flora and in some clinical specimens. An organism will be overlooked entirely if other organisms present in the mixture outnumber it significantly. Nevertheless, such "hidden" organisms may play important roles in pathological or other processes. Thus, for "total flora" studies, it is important to use appropriate selective media.

Selective media may permit earlier recovery and identification of organisms as well, a factor of considerable value in the processing of clinical specimens. Certain media are so selective that the presence of an organism on them, perhaps with the additional information provided by morphological features, provides reliable tentative identification.

Selective media are particularly useful in anaerobic work for two reasons:

(1) the vast majority of organisms isolated from humans which grow aerobically also grow well anaerobically, and

(2) anaerobes are commonly present in mixed culture (with both facultative organisms and other anaerobes).

Selective media would be valuable in the rumen roll-tube technique, as applied to study of human anaerobes. While the media to be described in this report would probably work well in roll-tubes, this has not been specifically studied.

Selective Media for Anaerobes

Table 1 lists the 6 selective media we have found most useful, together with the organisms for which they are most valuable. In Table 2, we indicate the organisms which will grow on the various media, other than those for which the media are used primarily.

I

TABLE 1. Selective media for anaerobes of human origin

Medium	Designation	Major uses
Neomycin blood agar	NEO	Anaerobic cocci (other than *Veillonella*) *Clostridium*
Kanamycin blood agar	KANA	*Bifidobacterium*
Kanamycin-vancomycin blood agar	KV	*Bacteroides fragilis*
Kanamycin-vancomycin laked blood agar	LKV	*Bacteroides melaninogenicus*
Neomycin-vancomycin blood agar	NV	*Sphaerophorus* (all species) *Fusobacterium fusiforme* *Veillonella*
Rifampin blood agar	RIF	*Sphaerophorus mortiferus* and *S. varius*

* The antibiotics were obtained from the sources shown in parentheses: Neomycin (The Upjohn Company, Kalamazoo, Mich.); Kanamycin (Bristol Laboratories, Syracuse, N.Y.); Vancomycin (Eli Lilly & Co., Indianapolis, Ind.), and Rifampin (Ciba Pharmaceutical Co., Summit, N.J.).

TABLE 2. Selective media for anaerobes of human origin—**unwanted** organisms which persist on various media

	NEO*	KANA*	KV*	LKV*	NV*	RIF*
Facultative:						
Gram-negative						
bacilli	+	+	+	+	+	
Streptococci	++++	++++				
Staphylococci	++++	++++				
Veillonella	++++	++++	++++	++++	@	
Gram-positive						
anaerobic and micro-						
aerophilic cocci	@	++	++	++	++	+
Gram-positive non-						
sporulating anaer-						
obic bacilli	++++	@	++	++	++	++
Clostridium	@	++++				+
Bacteroides	++++		@	@	++++	±
Fusobacterium	++++	+++	+++	+++	@	
Sphaerophorus	++++	+++	+++	+++	@	@

* For details, see Table 1; @, major use of medium, and ++++ to ± frequency and amount of growth. Absence of mark means no growth occurs.

The media are made as follows:

NEO—Neomycin blood agar

These are standard blood agar plates containing 100 μg/ml of neomycin base activity (0·5 g of neomycin sulfate is equal to 0·35 g neomycin base).

Add neomycin to the blood agar base prior to autoclaving as directed by the manufacturer of the blood agar base. Commercial vials of neomycin contain 0·5 g neomycin sulphate; add 1 ml sterile distilled water and mix well. Remove 0·3 ml and add to 1 litre of medium for proper final concentration.

Neomycin may also be used in egg yolk agar plates, Nagler plates, or together with sodium azide (Clostrisel agar, BBL) when one is looking particularly for *Clostridium*.

KANA—Kanamycin blood agar

These are standard blood agar plates containing 75 µg/ml of kanamycin base. Prepare as for neomycin blood agar. Kanamycin is available commercially in vials containing 1·0 g of kanamycin sulphate in solution (3 ml). Use 0·3 ml/litre of medium for proper final concentration.

KV—Kanamycin-vancomycin blood agar

These are standard blood agar plates containing 100 µg/ml of kanamycin base and 7·5 µg/ml of vancomycin. Use 0·4 ml of commercial kanamycin solution per litre of medium. Dissolve vancomycin hydrochloride in distilled water (pH 7·0), sterilize by passage through a 0·45 µ pore size membrane filter, and add aseptically (with the blood) to the previously autoclaved (standard autoclave procedure) and cooled blood agar base (7·5 mg/litre medium). If vancomycin **sulphate** is used, slight heating may be necessary to effect solution (swirl briefly in 45° water bath).

LKV—Kanamycin-vancomycin laked blood agar

This is similar to KV, but kanamycin is used in a concentration of 75 µg/ml, the blood is laked by freezing and thawing, and menadione is added. If commercial kanamycin solution is used, use 0·3 ml/litre of medium.

Make stock solution of menadione by dissolving 50 mg of the compound in 100 ml of 95% ethyl alcohol. Sterilize this solution by filtration through a membrane filter (0·45 µ pore size), or by autoclaving at 115° for 10 min. Refrigerate until needed. Add 1 ml of this stock solution aseptically to each litre of medium (**after** the medium has been autoclaved by the standard procedure), giving a final concentration of menadione of 0·5 µg/ml.

NV—Neomycin-vancomycin blood agar

This is made in exactly the same way as kanamycin-vancomycin blood agar except that if commercial neomycin sulphate is used, 1 ml water is added and 0·3 ml is removed and added to 1 litre of medium.

RIF—Rifampin blood agar

These are standard blood agar plates containing rifampin in a final concentration of $50\,\mu g/ml$; the drug must be added aseptically, together with the blood, after the blood agar base has been sterilized. Dissolve the rifampin in absolute ethyl alcohol and then add 4 volumes of distilled water. Make further dilutions with water. Stock solutions of $1000\,\mu g/ml$ may be kept in the refrigerator for up to 2 months. Plates are prepared the day they are to be used.

As with other media for anaerobic culture, it is desirable when feasible, to make these plates under anaerobic conditions (in an anaerobic chamber) and to store them anaerobically until time of use. All of these antibiotics with the exception of rifampin (see above), are stable at refrigerator temperatures for at least 1 month, although it is generally desirable to use relatively fresh plates.

Growth on Media

Examples of the usefulness of selective media for isolation of anaerobes are given in the accompanying illustrations. Figures 1–4 show platings of a single swabbing of healthy human gingival sulcus material on, respectively, blood agar (BAP), laked blood agar (LBAP), KV (KV BAP), and LKV (KV LBAP). It is apparent ,when one inspects the plate containing the appropriate selective (and differential) medium (LKV, Fig. 4) that *Bacteroides melaninogenicus* (note black colonies) is a major and perhaps dominant component of the normal flora of this region in the person studied. Yet very few black colonies are seen on the other plates. The KV plate, Fig. 3, shows sparser growth than the two nonselective media (Figs 1 and 2). However, the concentration of kanamycin in this medium (designed primarily for *Bacteroides fragilis*) is too high for some strains of *B. melaninogenicus*. Furthermore, the laked blood used in the selective medium for *B. melaninogenicus* (LKV) stimulates the growth of this organism and leads to much earlier development of the characteristic black pigment. Nevertheless, note (Fig. 2) that laked blood without the selecting agents (kanamycin and vancomycin) is not adequate for demonstration of *B. melaninogenicus*.

Figs 5–7 are platings of the same dilution of normal human faeces (10^{-6} dilution) on, respectively, blood agar (BAP), KV (KV BAP) and RIF (RIF BAP). Note that the KV has thinned out the population considerably, compared to the non-selective blood agar. Only obligate anaerobes, chiefly *B. fragilis*, are present on this plate. The RIF plate is dramatically selective, as illustrated, and comes close to being the ideal

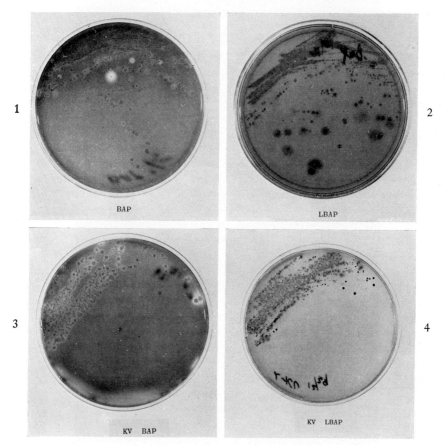

FIGS 1–4. A swab of the gingival sulcus area of a healthy adult has been streaked
on the four media depicted above. See text for description.

selective (for a single species) medium. Although organisms other than
Sphaerophorus mortiferus and *Sphaerophorus varius* sometimes grow on this
medium, the *Sphaerophorus* strains are readily identified by the large
translucent colonies with a central density (viewed by transmitted light) or
the "fried egg" appearance with central elevation (viewed by reflected
light). Another example of the excellent selective features of the RIF
medium is illustrated in Figs 8 and 9. This is another specimen from the
same person whose cultures are shown in Figs 5–7. The rifampin used in
the earlier example was somewhat outdated. In the two examples illustrated,
the presence of *Sphaerophorus* in the stool specimen could not be detected
except by use of RIF plates because the numbers present were relatively

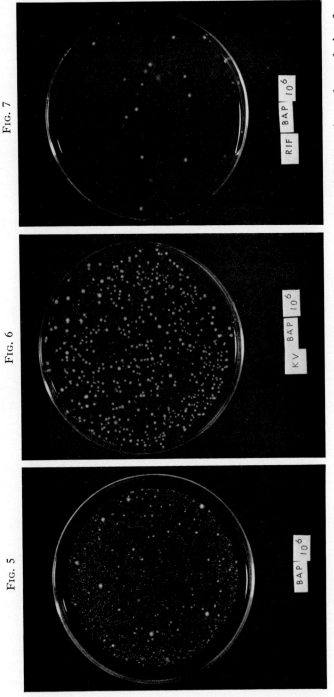

Fig. 5   Fig. 6   Fig. 7

Figs 5–7. A dilution (10^{-6}) of faecal specimen from a normal adult has been spread (0·5 ml/plate) on the surface of the 3 media depicted above. See text for description.

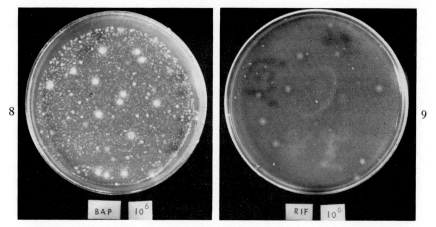

FIGS 8–9. A dilution (10⁻⁶) of faecal specimen, taken on another occasion from the same person whose cultures are pictured in Figs 5–7, plated (0·5 ml/plate) on above 2 media. See text for description.

low (c. 10^7/g wet faeces) and these organisms were obscured by the much larger number of others (total counts c. 10^{10}/g wet faeces).

Discussion

Various workers have used substances such as brilliant green, crystal or ethyl violet, bile, sodium azide, chloral hydrate, phenylethyl alcohol and antibiotics such as tyrothricin, streptomycin and polymyxin in selective media for anaerobic bacteria (Finegold, Miller and Posnick, 1965a). In our experience, the antibiotics reported on in the present paper have performed more efficiently and dependably. The background of susceptibility of various anaerobes to the antibiotics utilized which suggested their potential in selective media has been published previously (Finegold, 1959; Finegold et al., 1965a; Finegold et al., 1965c; Finegold et al., 1967; Finegold, Siewert and Hewitt, 1957; Finegold, Sutter and Sugihara, 1969; Miller and Finegold, 1967). Lowbury and Lilly (1955) suggested the use of neomycin in a selective medium for *Clostridium*. In 1957, this laboratory recommended neomycin, with or without vancomycin, for the isolation of Gram-negative non-sporulating anaerobic bacilli and other anaerobes (Finegold et al., 1957). Subsequently, Rogosa et al. (1958) adopted our use of 7·5 µg/ml of vancomycin for a selective medium for *Veillonella*. In 1959, we reported that kanamycin was even less active against *Bacteroides* than neomycin and that it might therefore prove more suitable in a selective medium for this genus (Finegold, 1959). Dowell, Hill and Altemeier (1962) recommended the use of menadione-kanamycin in media selective for anaerobes and Vera (1962) suggested the use of kanamycin discs on the surface of solid media to help select (within the zone of inhibition about these discs) anaerobes.

It is important to realize that the recommendations made herein apply only to anaerobes isolated from humans. Several studies have shown clearly that isolates of other origins (poultry, rumen) may have distinctly different antibiotic susceptibility patterns.

As indicated earlier, growth of organisms on certain selective media may provide helpful clues to rapid identification. Thus, black colonies on LKV are likely to represent *B. melaninogenicus*, large translucent colonies with dense centres on RIF are *S. mortiferus* or *S. varius*. Large, grey, entire smooth colonies on KV are probably *B. fragilis*. Speckled colonies with greening about them on NV are likely to be *Fusobacterium fusiforme*. Spreading colonies on NEO are usually *Clostridium*. The problem of facultative organisms growing anaerobically has already been alluded to. The 3 media containing an aminoglycoside plus vancomycin (KV, LKV, and NV) and RIF are very valuable in this regard. Growth on any of these 4 media is good presumptive evidence of an obligate anaerobe as facultative organisms seldom grow on these media.

We have already mentioned that organisms present in significant numbers may be undetectable in the presence of larger numbers of other organisms. This was well illustrated in the text in the case of *Sphaerophorus*. It is interesting to note that about half of normal adults can be shown to have *Sphaerophorus* in the faeces and that mean counts of this organism, when present, are 10^9/g wet faeces (Finegold and Miller, 1968; Finegold *et al.*, 1965*b*). It does not seem reasonable that the other half of the population has no *Sphaerophorus* at all. Our demonstration, in this paper, of an adult with 10^7 *Sphaerophorus*/g wet faeces, detectable only with the RIF medium, suggests the likelihood that surveys with appropriate selective media would demonstrate smaller numbers of *Sphaerophorus* in some of the people not showing them in studies not utilizing selective media. Similarly, *Bifidobacterium* was found (Finegold and Miller, 1968; Finegold *et al.*, 1965*a*) in the faeces of two-thirds of adults, with mean counts (when present) of 10^9. Again here, it seems likely that appropriate selective media would reveal smaller numbers of bifids in some of the other people.

Further modifications could be made to improve the selectivity of some of the media. Thus, the addition of penicillin (1 μg/ml) or of polymyxin B (12·5 μg/ml) would eliminate a number of organisms other than *B. fragilis* which can grow on KV. On the other hand, as implied earlier, the morphology of *B. fragilis* is usually distinctive enough to allow one to recognize it on this medium. Erythromycin, added to NV in a concentration of 25 μg/ml, would very likely make this medium much more selective for *Sphaerophorus* spp. While the modifications suggested in this paragraph have not been specifically studied, plate dilution antibiotic susceptibility data (Finegold *et al.*, 1967) suggest that they should perform as indicated. Other types of additives might also be useful. Thus, the use of 20% bile, though unreliable as an identification test in blood agar, might serve a purpose similar to that proposed above for penicillin or polymyxin B in KV and would also serve to stimulate the growth of many strains of *B. fragilis* (Shimada and Finegold, 1969). Desoxycholate agar (BBL) (Shimada and Finegold, 1969), could be used, with the addition of neomycin (100 μg/ml), as a selective medium for several species of *Sphaerophorus* (*S. mortiferus*, *S. varius*, and *S. ridiculosus*), assuming that there was no antagonism between the components.

Nalidixic acid, in a concentration of 100 μg/ml, should theoretically be useful in a selective medium for anaerobic cocci and for *Bifidobacterium*, although a number of other organisms (*Veillonella*, Gram-negative anaerobic bacilli, some clostridia, *Pseudomonas* and occasional strains of other aerobic Gram-negative bacilli, streptococci and some staphylococci) would also be expected to grow on this medium. Preliminary experience with a medium of this type does not suggest that it would offer an advantage over the other media (NEO and KANA) proposed for these two organisms.

Still another approach to improving selectivity of media is to use less than optimum anaerobic conditions to facilitate recovery of less fastidious organisms such as *Clostridium perfringens*. Thus it may be that the high degree of selectivity for this organism claimed by some workers for neomycin blood agar was in large part related to the use of relatively poor anaerobiosis such that other anaerobes tolerating neomycin (viz., the non-sporulating bacilli) could not grow.

Much additional work is needed to further improve presently available selective media and to provide additional media. A selective medium for the anaerobic members of the *Lactobacilleae* is badly needed. The KANA medium proposed for *Bifidobacterium* is poor because so many other anaerobes and aerobes grow on it. Nothing is available for other *Lactobacilleae*. These organisms are often present in the intestinal tract, but are commonly overlooked. Additional media are needed for other members of the normal oral flora. As indicated earlier, use of selective media in the roll-tube technique would certainly facilitate utilization of this excellent technique.

The antibiotics found useful in solid selective media may also be utilized effectively in liquid media, particularly for enrichment purposes.

Finally, it is always important to bear in mind that selective media should always be used together with non-selective media. One must always verify the identity of organisms recovered from selective media by means of appropriate tests.

Summary

Information concerned with 6 selective media found very useful for isolation of various anaerobic bacteria of human origin is presented along with suggestions for certain modifications of media or technique which may lead to improved results. These media are very useful in studies of normal anaerobic flora of the body and in processing clinical specimens which may contain anaerobes.

References

DOWELL, JR., V. R., HILL, E. O. & ALTEMEIER, W. A. (1962). Methods for the isolation and identification of non-sporulating anaerobic bacteria from clinical specimens. *Bact. Proc.,* 90.

FINEGOLD, S. M. (1959). Kanamycin. *Archs intern. Med.,* **104,** 15.

FINEGOLD, S. M. & MILLER, L. G. (1968). Normal fecal flora of adult humans. *Bact. Proc.,* 93.

FINEGOLD, S. M., MILLER, A. B. & POSNICK, D. J. (1965*a*). Further studies on selective media for *Bacteroides* and other anaerobes. International mikroökologisches Symposium, Potsdam, Germany, 1964, *Ernahrungsforschung,* **10,** 517.

FINEGOLD, S. M., POSNICK, D. J., MILLER, L. G. & HEWITT, W. L. (1965*b*). The effect of various antibacterial compounds on the normal human fecal flora. International mikroökologisches Symposium, Potsdam, Germany, 1965, *Ernährungsforschung,* **10,** 316.

FINEGOLD, S. M., POSNICK, D. J., MILLER, L. G. & MILLER, A. B. (1965*c*). Sensitivity of major *Bacteroides* groups to antibacterial agents. *Bact. Proc.,* 64.

FINEGOLD, S. M., POSNICK, D. J., MILLER, L. G. & MILLER, A. B. (1967). Antibiotic susceptibility patterns as aids in classification and character-ization of Gram-negative anaerobic bacilli. *J. Bact.,* **94,** 1443.

FINEGOLD, S. M., SIEWERT, L. A. & HEWITT, W. L. (1957). Simple selective media for *Bacteroides* and other anaerobes. *Bact. Proc.,* 59.

FINEGOLD, S. M., SUTTER, V. L. & SUGIHARA, P. T. (1969). Suscept-ibility of anaerobic bacteria to rifampicin. *Bact. Proc.,* 73.

LOWBURY, E. J. L. & LILLY, H. A. (1955). A selective plate medium for *Cl. welchii. J. Path. Bact.,* **70,** 105.

MILLER, L. G. & FINEGOLD, S. M. (1967). Antibacterial sensitivity of *Bifidobacterium (Lactobacillus bifidus), J. Bact.,* **93,** 125.

ROGOSA, M. FITZGERALD, R. J., MacKINTOSH, M. E. & BEAMAN, A. J. (1958). Improved medium for selective isolation of *Veillonella. J. Bact.,* **76,** 455.

SHIMADA, K. & FINEGOLD, S. M. (1969). Effect of bile and desoxycholate on Gram-negative anaerobic bacteria. *Bact. Proc.,* 87.

VERA, H. D. (1962). Bacteriology aids suggested—Dr. Harriette D. Vera gives advice on media problems. *Lab. World,* **13,** 1084.

Isolation of *Bacteroides fragilis* and *Sphaerophorus-Fusiformis* Groups

H. Beerens and L. Fievez

Institut Pasteur, Lille, France and Faculté de Médecine Vétérinaire, Bruxelles Cureghem, Belgium

Amongst the Gram-negative anaerobic bacteria two groups of organisms are of particular importance in man and animals, the *Bacteroides fragilis* group and the *Sphaerophorus-Fusiformis* group.

Bacteroides fragilis

Bacteroides of this group (Eggerth and Gagnon, 1933) are represented by several species of which the type species is *B. fragilis*. *B. vulgatus* and *B. thetaiotaomicron* are also recognized. Certain strains are pathogenic and are responsible for infections which may be severe such as septicaemia, meningitis, and multiple suppurations. Others are saprophytes, the intestine without doubt being the most common habitat of these species. There are several methods for isolating these organisms.

From pathological material

1. *When the* Bacteroides *are present as a relatively pure culture, isolation in deep VL medium is recommended.*

Formula of Medium

Tryptone	10 g
NaCl	5 g
Meat extract	2 g
Yeast extract	5 g
Cysteine hydrochloride	0·3 g
Glucose	2 g
Agar (Difco)	6 g
Distilled water	1000 ml

Adjust to pH 7·4. Distribute in 180 × 19 mm tubes in 5 ml amounts.

Sterilize by autoclaving at 115° for 20 min. The addition of 10% beef bile encourages the growth of a very large number of *Bacteroides*.

Inoculation

With a closed Pasteur pipette take a sample of a portion of the pus or the pathological material and add the inoculum to the 1st tube of medium, previously melted in a water bath (100°) and cooled to 45°. Add the inoculum with the closed Pasteur pipette to the 1st tube taking care not to aerate the medium by too rapid movements, continue by introducing the same pipette into the 2nd tube, then the 3rd tube, etc. Cool under running cold water. Incubate at 37°.

Results

Ordinary VL medium. Punctiform colonies appear in 48 h. The organisms are strict anaerobes.

VL medium with 10% bile

The colonies appear in 24 h. They are very large and surrounded by a whitish precipitate of bile salts produced by reaction with the acid from the fermentation of the glucose.

2. *When* Bacteroides *are associated with other anaerobic bacteria*

Streak the material in parallel lines on the surface of the VL medium containing blood or VL bile sodium azide medium as described in 3.

Formula of VL medium with blood

Tryptone	10 g
NaCl	5 g
Meat extract	2 g
Yeast extract	5 g
Cysteine hydrochloride	0·3 g
Glucose	2 g
Agar (Difco)	20 g
Distilled water	1000 ml

Adjust to pH 7·4, distribute in 20 ml amounts per tube. Sterilize at 115° for 20 min. When required melt in a boiling water bath and cool to 46–48°, add 2·5 ml sterile horse blood. Pour into Petri dishes. Streak out, incubate anaerobically for 48 h. The colonies of *Bacteroides* are translucent and 1 mm in diameter.

3. *When* Bacteroides *are associated with aerobes in particular*

Enterobacteriaceae

(a) Inoculate 4 or 5 drops of sample into a tube of liquid anaerobic medium

(Rosenow's Medium (see Buttiaux, Beerens and Tacquet, 1969) or Brain Heart (Difco) medium). Incubate anaerobically for 48 h.

(b) Inoculate this culture on VL medium containing bile and sodium azide.

Formula of medium

Tryptone	10 g
NaCl	5 g
Meat extract	2 g
Yeast extract	5 g
Cysteine hydrochloride	0·3 g
Glucose	2 g
Agar (Difco)	20 g
Distilled water	1000 ml

Results

The colonies are round (1 mm in diameter), translucent and are found at the highest dilutions. The *Enterobacteriaceae* and faecal streptococci only occur in the lower dilutions so no longer appear on the plates.

Sphaerophorus-Fusiformis

Sphaerophorus can be found in pathological and faecal material. There are several methods of isolation.

From pathological material

1. *When the bacteria exist in pure culture*

The procedure for isolation is identical to that described for the *Bacteroides fragilis* group.

2. *When* Sphaerophorus *strains are associated with other anaerobic bacteria*

The procedures recommended are either isolation on the surface of VL blood agar plates (as described for the *Bacteroides fragilis* group), or the utilization of selective medium described in 3 below.

3. *When* Sphaerophorus *strains are associated with aerobic bacteria, in particular* Enterobacteriaceae *as for example in specimens of faecal material*

The use of a medium containing brilliant green and sodium azide is recommended. The formula is as follows.

Tryptone	20 g
Sodium chloride	10 g
Meat extract	6 g

Yeast extract	10 g
Cysteine hydrochloride	0·8 g
Glucose	4 g
Agar (Difco)	40 g
Distilled water	1000 ml

Adjust the pH to 7·2. Distribute in bottles and sterilize at 120° for 20 min. When required, melt in a boiling water bath, cool to 46–48° and add per 1000 ml of medium, 500 ml of solution 1, and 500 ml of solution 2.

Solution 1.
Brilliant green, 0·072 g
Distilled water, 1000 ml

Solution 2.
Sodium azide, 1·20 g
Distilled water, 1000 ml

Adjust to pH 7·4. Distribute in 20 ml per tube. Sterilize at 115° for 20 min. When required, melt in a boiling water bath and cool to 46–48°.

Add 2·5 ml sterile horse blood and 4·5 ml of a solution of bile + Sodium azide (sterile beef bile 100 ml, sodium azide 0·05 g). Mix gently without aeration and pour into Petri dishes.

With a platinum wire take a sample from the liquid culture and streak on the surface of the medium (incubate at 37°, anaerobically.)

Results

The colonies appear in 48 h, they are translucent and punctiform.

From faecal material

There are 2 considerations: the isolation and the enumeration (Buttiaux, *et al.*, 1969).

1. *Isolation*

Use the procedure described in 3 above.

2. *Enumeration*

Use modified medium VL, semi-solid.

Tryptone	10 g
Sodium chloride	5 g
Meat extract	2 g
Yeast extract	5 g
Cysteine hydrochloride	0·3 g
Glucose	2 g
Agar (Difco)	0·5 g
Distilled water	1000 ml

Adjust the pH to 7·4. Distribute (9 ml amounts) in 160 × 16 mm tubes. Sterilize at 115° for 20 min. When required, reheat by holding in a boiling water bath for 20 min and cool to 45–48°.

Homogenize 1 g of faecal material in a tube containing 9 ml of medium. Distribute the homogenate in the medium with a closed Pasteur pipette without introducing any air. Make decimal dilutions by taking 1 ml of this first dilution (10^{-1}) which is then transferred into a second tube containing 9 ml (10^{-2}). Distribute the inoculum as previously, with a closed Pasteur pipette, taking a 1 ml sample of the second dilution which is then transferred to the third tube (10^{-3}) containing 9 ml of medium etc., up to 10^{-10}. Inoculate 0·1 ml of each dilution on to the surface of VL medium containing blood previously poured into Petri dishes as described in 2 above. Incubate anaerobically at 37° for 48 h. Adjust the final pH of the medium to 7·0, then add 3% horse blood. Pour into Petri dishes.

Inoculation. This is done either directly from the material to be examined, diluted *c.* 1/5 in VL semi-solid medium as described for *Bacteroides fragilis* in the previous section, or from preliminary enrichment of the material, in Rosenow's medium or Brain heart (Difco) medium. Incubate for 48 h anaerobically. In this case take a loopful of the culture and streak out to obtain isolated colonies on the surface of the medium. In both cases incubate anaerobically for 48 h.

Results. Colonies of *Sphaerophorus necrophorus* are surrounded by a zone of haemolysis and they exhibit a faint yellowish centre. After prolonged incubation for 3 days, the periphery of the colony exhibits a halo with a metallic appearance.

Sphaerophorus funduliformis or *Sphaerophorus pseudonecrophorus* appear as opaque colonies slightly convex with regular edges 2 mm diam. They are sometimes yellowish and may or may not be haemolytic.

Most of the saprophytic *Fusiformis* strains are cultivated on this medium. The pathogenic *Fusiformis* strains isolated for example from pleural suppurations require a medium with serum or blood (VL containing blood as described for *Bacteroides fragilis* group on p. 109) but cannot be cultivated on the selective medium containing brilliant green and sodium azide.

References

BUTTIAUX, R., BEERENS, H. & TACQUET, A. (1969). *Manual of Bacteriological Techniques.* 3rd Edition. Paris: Flammarion Medicales.

EGGERTH, A. H. & GAGNON, B. H. (1933). The bacteroides of human faeces. *J. Bact.,* **25,** 389.

The Isolation of the Anaerobic Bacteria from Chicken Caeca with particular reference to members of the Family *Bacteroidaceae*

ELLA M. BARNES AND C. S. IMPEY

Food Research Institute, Norwich, England

The anaerobes form a major part of the caecal flora of chickens, occurring at more than 10^{10}/g (Ochi, Mitsuoka and Sega, 1964). With 5-week-old chickens fed on a normal broiler diet, Barnes and Impey (1970) found that the Gram-negative non-sporing rods (*Bacteroidaceae*) formed about 40% of the population and were equalled in numbers by the Gram-positive non-sporing rods (including bifidobacteria) while peptostreptococci occurred at about 15%. Curved rods (possibly spirilla) were also isolated together with other unidentified anaerobes. A number of different groups of Gram-negative non-sporing rods isolated from poultry have now been described. Of these, some have so far only been obtained when non-selective media were used but for others selective media have been developed (Barnes and Goldberg, 1962; Goldberg, Barnes and Charles, 1964; Barnes and Impey, 1968, 1970).

The main problem in the isolation of anaerobes, is the establishment and maintenance of a suitable anaerobic environment whilst collecting and diluting the sample and growing the organisms. Two different techniques are frequently used. One is the strict anaerobic roll-tube technique originally described by Hungate (1950) for the isolation of the rumen microorganisms. The other is the traditional anaerobic technique where the plates are incubated in an anaerobic jar.

In the tests carried out by Barnes and Impey (1970) it was found that the viable count in the Hungate roll-tubes was about four times that obtained using plates incubated in the anaerobic jar, although there is no evidence at present that any of the chicken strains can only be grown in pure culture by the Hungate technique. There is a considerable difference, however, between the initiation of growth from a large inoculum under less favourable redox conditions and the multiplication of single or damaged cells under similar conditions. The great advantage that surface inoculation of anaerobic plates has over roll-tubes is that it is much easier to observe different colony types and purify mixed cultures. However it is well known that there are a number of very strict anaerobes such as the cellulolytic and methane-producing bacteria which may only be recovered by using a technique similar to that of Hungate. Although these organisms have not been found to occur in significant numbers in the chicken fed on a normal broiler diet, they may occur in much larger numbers in other birds or when different diets are used. Shrimpton (1966) found a much higher concentration of methane in the gases obtained from the caeca of fistulated chickens fed on a diet with a higher percentage of crude fibre.

The recommended isolation procedure for the chicken caecal flora is described below. The dilutions are prepared using the Hungate technique. These are then used to inoculate roll-tubes and at the same time are plated out on pre-reduced non-selective and selective agar plates which are incubated in the anaerobic jar. The same dilutions may also be used for isolating the facultative anaerobes such as faecal streptococci, lactobacilli and enterobacteria.

Methods

Anaerobic techniques

Hungate roll-tube technique

This technique has been described and discussed by Hungate (1966) and the method has been modified by Moore (1966) to cope with large numbers of samples. In principle the technique relies on the exclusion of all traces of oxygen both in the preparation of the medium and in the dilution of the sample. This is accomplished by carrying out all operations under a continual flow of nitrogen or carbon dioxide from which the last traces of oxygen have been removed by passing over heated copper. The column used is similar to that described by Moore (1966)*. The diluents and all the media contain resazurin which should stay colourless at every stage during the dilution of the sample and preparation of the roll-tubes.

Traditional anaerobic technique

In using the traditional anaerobic techniques the following precautions must be taken.

Growth on agar plates. All the plates used for the isolation or purification of the anaerobes should be poured several days before use and stored in the anaerobic jar. This ensures that the agar plates are in a reduced condition when inoculated. A mixture of hydrogen and 10% carbon dioxide should

* See also p. 134.

be used in the jar which must also contain a catalyst to ensure the removal of the last traces of oxygen. Plates must be placed in the anaerobic jar within 5 min of inoculation. A tube of methylene blue anaerobic indicator solution is always included in the jar (Cruickshank, 1962 and this book, pp. 66, 83).

Growth in liquid media. This can be carried out without using an anaerobic jar providing that the following precautions are taken.

1. A semi-solid reducing medium should be used such as RCM or VL to which supplements may be added.
2. It should be distributed in tubes or small bottles which can be tightly sealed.
3. Only a small head space should be left in the bottle.
4. The medium should be held in a boiling water bath for at least 20 min to remove oxygen, then cooled and inoculated immediately.
5. A large inoculum (*c.* 0·25 ml) should be used.

Media

For use with the Hungate techniques

Anaerobic dilution solution (ADS), Bryant and Burkey (1953).

Supplemented Medium 10

The medium used is the Medium 10 of Caldwell and Bryant (1966) which has been supplemented with liver extract 5% and chicken faecal extract 10%. The extracts are incorporated in the basal medium before autoclaving, adjustment being made for the required amount of water.

Liver extract

Dehydrated liver (Difco) 27 g are dissolved in 200 ml distilled water, heated to 50° and held at this temperature for 1 h. The mixture is then boiled, cooled and centrifuged. The pH of the supernatant is adjusted to 7·0–7·2 and sterilized by autoclaving (121°/15 min).

Faecal extract

This is prepared by autoclaving (121°/30 min) equal quantities of chicken faeces and water. The sludge is centrifuged and the supernatant poured off and adjusted to pH 7·0–7·2. It is sterilized by autoclaving (121°/15 min).

For use with the traditional anaerobic techniques

VL medium

Modified from Beerens *et al.* (1963) contains (g/1): Tryptone (Oxoid), 10; NaCl, 5; beef extract (Lab-Lemco, Oxoid), 3; yeast extract (Difco), 5; cysteine hydrochloride, 0·4; glucose, 2·5; agar (New Zealand), 0·6; pH 7·2–7·4. The medium is sterilized (121°/15 min) in 19 ml lots in 1 oz screw-capped bottles.

For the solid medium, agar (New Zealand) 12 g/1 is added.

VLBM agar

VL agar containing laked blood 5% and menadione 0·5 μg/ml.

Laked blood

Prepared from oxalated horse blood (Wellcome Reagents Ltd.) by freezing and thawing three times.

Menadione

A menadione (Sigma Chemical Co.) solution 200 μg/ml is prepared by dissolving 20 mg in 2 ml ethyl alcohol, diluting to 100 ml and sterilizing by filtration through a millipore filter. The solution must be held in a refrigerator.

VL haemin broth or agar

VL medium containing haemin (Koch Light Laboratories) 1 μg/ml.

Haemin solution

A 40 μg/ml solution is prepared by dissolving 10 mg haemin in 1 ml of nNaOH and making the volume up to 250 ml. Sterilize by autoclaving 121°/15 min.

VLBM kanamycin agar

Add kanamycin sulphate (The Bayer Products Co.) 100 μg/ml to the VLBM agar immediately before pouring the plates.

BGP

VL medium containing 0·1% glucose instead of 0·25%, together with 0·4% Na_2HPO_4. It is used with or without supplements for the maintenance of cultures.

RCM

This medium is prepared according to the original formula of Hirsch and Grinsted (1954).

Ethyl violet azide agar. (EV Az)

RCM agar containing ethyl violet 1/20,000 and sodium azide 1/20,000 (Barnes and Goldberg, 1962).

Isolation procedures

Preparation of the sample

The birds are killed by the administration of nembutol and the caecal samples are taken for microbiological examination without delay. All of the following operations are then carried out under a continuous flow of oxygen-free carbon dioxide. The contents of the caecum are squeezed into a weighed tube containing 9 ml of anaerobic dilution solution (ADS) and some glass beads. The tube is shaken and ten-fold dilutions are prepared using the same diluent. The dilutions may be used as described below.

Direct microscopical count

It is essential to perform a direct microscopical count on every caecal sample analysed. It is only by doing this that one has a reference point to assess the efficiency of the isolation procedures used. The caecal suspension is diluted to the required concentration (generally 1/10) in saline (0·9% w/v) containing formalin 10%. The organisms are counted in a Helber counting chamber using the phase contrast microscope and following the method described by Meynell and Meynell (1965).

Isolation of anaerobes using the Hungate roll-tube technique

This technique is used to determine the maximum numbers and types of anaerobes present. A portion (1 ml) of each of the required dilutions is added in duplicate or triplicate to 9 ml of pre-reduced supplemented Medium 10 (see above) and a roll-tube prepared. The tubes are incubated for about 1 week at 37° or until no more colonies develop. It was found by Barnes and Impey (1970) that in order to recover the maximum numbers of anaerobes from the chicken caecum it was necessary to add liver and faecal extract to the Medium 10. The use of faecal extract was suggested by Dr T. Gibson who had found it stimulatory in similar experiments, whilst the authors had found liver extract to be a useful growth supplement for certain of the Gram-negative anaerobes.

To isolate the organisms, the colonies are picked from the roll-tubes using a stout platinum or steel wire and purified by streaking up a roll-tube of the same medium. Alternatively if there are only a few colonies in the tube they can be stabbed directly into a supplemented Medium 10 slant. Growth generally occurs in 1–3 days. Organisms in the water of syneresis from cultures showing growth are examined in wet mounts by phase contrast microscopy for morphology, motility and homogeneity and Gram stained. At this point the organisms are also tested for their ability to grow on VLBM agar using the traditional anaerobic technique. If the organism grows well on the agar plate one can then either continue using the Hungate technique for further tests or proceed by the method described below. If the organism fails to grow it is one of the stricter anaerobes possibly resembling those isolated from the rumen and for its characterization reference should be made to Bryant (1963).

The roll-tube technique may also be used for the isolation of particular groups of organisms such as those which utilize starch. In this latter case the carbohydrates in Medium 10 are replaced by starch (0·5%). Cellulolytic strains are isolated by incorporating 0·2% pebble mill ground Whatman No. 1 filter paper in place of the carbohydrates or alternatively using the method and medium described by Mann (1968).

Isolation of anaerobes using the traditional anaerobic techniques

Total anaerobic count. Two drops (0·05 ml) of the required dilution are spread rapidly over the surface of half of a pre-reduced VLBM agar plate, taking the precautions described above. This medium has been found to be the most satisfactory for isolating and purifying the anaerobes. If whole blood is used for purifying the isolates instead of laked blood it has been found that many of them cannot be recovered. The reason for this is not yet known. Once a pure culture has been obtained it is subcultured into VL laked blood-menadione broth which is then used for a number of initial screening tests and for the preparation of a stock culture in BGP broth supplemented with laked blood and menadione. It is difficult to maintain many of these cultures for longer than 2 weeks without sub-culturing them so they should be freeze-dried as soon as possible.

The VLBM broth culture is examined for (a) aerobic growth to eliminate facultative anaerobes, (b) morphological appearance using phase contrast, a photograph being taken if possible, (c) Gram reaction, (d) gas and terminal pH and (e) volatile fatty acids following the method of Moore, Cato and Holdeman (1966). At this stage the organisms are also tested to determine whether they can be maintained without the supplements of laked blood and menadione in the medium, three transfers being carried out to ensure that there is no carryover of growth factors. Of the poultry anaerobes isolated so far, some will grow in the unsupplemented medium, others have a requirement for haemin whilst others cannot be grown without a small addition of blood (0·5%) which is added to the VL broth before it is held in the boiling water bath to remove the oxygen. The different growth requirements are shown in Table 1.

TABLE 1. Supplements required to support the growth of some of the chicken caecal anaerobes in VL broth

Organism	Supplements required for growth in the VL broth
Gram-negative non-sporing rods	
Group 1* (Unidentified)	None
4* (*B.hypermegas*)	None
5* (*B.fragilis* and related strains)	Haemin 1 μg/ml
7† (Unidentified)	Haemin 1 μg/ml
8† (Unidentified)	None
Peptostreptococci†	
Group 1 (Unidentified)	None
2 (Unidentified)	None
3 (Unidentified)	None
4 (Unidentified)	Laked blood 0·5 %
Gram-positive non-sporing rods†	
Group 1 (Bifidobacteria)	
2 (Unidentified)	Laked blood 0·5 %
3 (Unidentified)	

* Described by Barnes and Impey (1968). † Described by Barnes and Impey(1970).

Isolation of the Gram-negative non-sporing anaerobes

The Gram-negative non-sporing anaerobes may be isolated from the Hungate roll-tubes or from the VLBM agar plates providing that they form a significant proportion of the total anaerobic flora. In many cases, however, individual strains occur in smaller numbers than this and selective methods must be used. In certain cases as with some of the unidentified poultry strains (such as groups 7 and 8 of Barnes and Impey, 1970), none of the selective media have so far proved useful and reliance must at present be placed on isolation from the non-selective media. The recommended isolation media are shown in Table 2. For certain strains, particularly those resembling *Bacteroides fragilis*, kanamycin 100 μg/ml is added to the VLBM agar. The use of various combinations of antibiotics for the selective isolation of these organisms and others such as *Bacteroides melaninogenicus* strains is discussed by Finegold, Sugihara and Sutter (1970).

One of the most important groups of anaerobic Gram-negative rods in the caeca of chickens, turkeys and ducks, consists of the very large "group 4" organisms originally described by Goldberg, Barnes and Charles (1964) and now considered to be similar to *Bacteroides hypermegas* described by Harrison and Hansen (1963). These may be isolated using ethyl violet azide agar. A number of strains do not grow well on this medium but any reduction in the concentration of ethyl violet or sodium azide destroys the

TABLE 2. Isolation media for the Gram-negative non-sporing anaerobes (Bacteroidaceae) of chickens

Organism	Medium
All groups	Supplemented Medium 10 (roll-tubes) or VLBM agar
Group 1* (unidentified)	EV Az agar or VLBM kanamycin agar
Group 4* (*B.hypermegas*)	EV Az agar
Group 5* (*B. fragilis* and related strains)	VLBM kanamycin agar
Group 7† (unidentified)	VLBM agar
Group 8† (unidentified)	VLBM agar and possibly VLBM kanamycin agar

* Described by Barnes and Impey, 1968. † Described by Barnes and Impey, 1970.

selectivity of the medium (Barnes and Goldberg, 1962). These organisms are inhibited by neomycin 100 μg/ml or kanamycin 100 μg/ml so cannot be isolated on the media used for isolating the *B. fragilis* and related strains.

Ethanol producing Gram-negative anaerobes assigned to group 1 (Barnes and Impey, 1968) have been found occasionally at low levels in the chicken caecum. These grow well both in the presence of ethyl violet and kanamycin.

Discussion

In order to determine the ecological niche of any particular group of organisms in the caecum and the environmental factors which may lead to a change in the balance of organisms there, it is essential to develop selective media for each group, so that the laborious isolation and identification procedures which are necessary at present may be avoided. This can only be done when the main types present have been isolated and their particular physiological properties determined. The Gram-negative non-sporing anaerobes are now being studied in this way, but apart from the bifidobacteria, little is known of the Gram-positive non-sporing anaerobic rods and the peptostreptococci from the chicken caecum. There are also other organisms which have still never been isolated in pure culture.

No mention has been made here of the clostridia. They are not generally present in numbers greater than 10^6/g and may be isolated by any of the methods described in other chapters in this book.

References

BARNES, E. M. & GOLDBERG, H. (1962). The isolation of anaerobic Gram-negative bacteria from poultry reared with and without antiobiotic supplements. *J. appl. Bact.*, **25**, 94.

BARNES, E. M. & IMPEY, C. S. (1968). Anaerobic Gram-negative non-sporing bacteria from the caeca of poultry. *J. appl. Bact.*, **31**. 530.

BARNES, E. M. & IMPEY, C. S. (1970). The isolation and properties of the predominant anaerobic bacteria in the caeca of chickens and turkeys. *Br. Poult. Sci.* **11**, 467.

BEERENS, H., SCHAFFNER, Y., GUILLAUME, J. & CASTEL, M. M. (1963). Les bacilles anaérobies non sporulés à Gram-négatif favorisés par la bile. *Annls Inst. Pasteur Lille,* **14,** 5.

BRYANT, M. P. (1963). Symposium on microbial digestion in ruminants: identification of groups of anaerobic bacteria active in the rumen. *J. Anim. Sci.,* **22,** 801.

BRYANT, M. P. & BURKEY, L. A. (1953). Cultural methods and some characteristics of some of the more numerous groups of bacteria in the bovine rumen. *J. Dairy Sci.,* **36,** 205.

CALDWELL, D. R. & BRYANT, M. P. (1966). Medium without rumen fluid for non-selective enumeration and isolation of rumen bacteria. *Appl. Microbiol.,* **14,** 794.

CRUICKSHANK, R. (1962). *Mackie & McCartney's Handbook of Bacteriology,* Edinburgh and London: E. & S. Livingstone Ltd.

FINEGOLD, S. M., SUGIHARA, P. T. & SUTTER, V. L. (1971). Use of selective media for isolation of anaerobes from the human intestine. (This volume p. 99).

GOLDBERG, H. S., BARNES, E. M. & CHARLES, A. B. (1964). Unusual bacteroides like organism. *J. Bact.,* **87,** 737.

HARRISON, A. P., JR. & HANSEN, P. A. (1963). *Bacteroides hypermegas* nov. spec. *Antonie van Leeuwenhoek, 29,* 22.

HIRSCH, A. & GRINSTED, E. (1954). Methods for the growth and enumeration of anaerobic spore formers from cheese with observations on the effect of nisin. *J. Dairy Res.,* **21,** 101.

HUNGATE, R. E. (1950). The anaerobic mesophilic cellulolytic bacteria. *Bact. Rev.,* **14,** 1.

HUNGATE, R. E. (1966). *The rumen and its microbes.* New York and London: Academic Press.

MANN, S. O. (1968). An improved method for determining cellulolytic activity in anaerobic bacteria. *J. appl. Bact.,* **31,** 241.

MEYNELL, G. G. & MEYNELL, E. (1965). *Theory and practice in experimental bacteriology.* Cambridge: University Press.

MOORE, W. E. C. (1966). Techniques for routine culture of fastidious anaerobes. *Int. J. syst. Bact.,* **16,** 173.

MOORE, W. E. C., CATO, E. P. & HOLDEMAN, L. V. (1966). Fermentation patterns of some *Clostridium* species. *Int. J. syst. Bact.,* **16,** 383.

OCHI, Y., MITSUOKA, T. & SEGA, T. (1964). Untersuchungen über die Darmflora des Huhnes III Mitteilung: Die Entwicklung der Darmflora von Kuken bis zum Huhn. *Zentbl. Bakt. ParasitKde,* Abt I, **193,** 80.

SHRIMPTON, D. H. (1966). Metabolism of the intestinal microflora in birds and its possible influence on the composition of flavour precursors in their muscles. *J. appl. Bact.,* **29,** 222.

Isolation of Cellulolytic Bacteria from the Large Intestine of the Horse

M. ELIZABETH DAVIES

Royal (Dick) School of Veterinary Studies, Edinburgh, Scotland

The method here described for isolation of cellulolytic bacteria from the horse has been modified from those used on the ruminant (Davies, 1964). The sites of cellulose digestion in the horse are the caecum and the dorsal and ventral colon, and while conditions obtaining in these organs differ from those in the rumen, similar precautions for strict anaerobiosis are required.

Collection of Samples

These are obtained either from fistulated animals or at slaughter. They should be collected in screw-capped containers, filled to capacity, and containing a diluent to give a final concentration of 10 g contents/100 g suspension. The diluent consists of: NaCl, 6·0 g; KH_2PO_4, 1·0 g; $MgSO_4$. $7H_2O$, 0·1 g; anhydrous $CaCl_2$, 0·1 g; $NaHCO_3$, 0·6 g and distilled water to 1000 ml. The solution need not be sterilized if used immediately. Otherwise, steaming at 100°/1 h has been found adequate.

For satisfactory isolations the samples must be cultured as soon as possible and not later than 3 h after withdrawal. All procedures should be carried out in an atmosphere of carbon dioxide from which all traces of oxygen have been removed. The latter can be obtained by passing the CO_2 over a heated coil such as is found in the Sunvic TC11 furnace, (Cattison Smith Ltd., Adam Bridge Works, Wembley) at a temperature of 800°. Collection vessels for the samples and all other containers should be flushed with O_2-free CO_2 before use.

Medium

Base

A modification of the salt solution of Hungate (1950) for the isolation of cellulolytic bacteria from the rumen is used, and has the following composition: KH_2PO_4, 0·02%; K_2HPO_4, 0·03%; $MgSO_4$. $7H_2O$, 0·01%; $CaCl_2$ (anhydrous), 0·01%; NaCl, 0·1%; $(NH_4)_2SO_4$, 0·1%; $NaHCO_3$, 0·5% L-cysteine hydrochloride, 0·02%, resazurin 0·0001% in distilled water. This solution is heat-labile and is sterilized through a porcelain filter candle. It should be colourless or have the blue colour of reduced resazurin.

As the complete medium is most frequently used in solid form the base is usually prepared in double strength and rendered solid by the addition of an aqueous solution of 3·0% (w/v) agar, previously sterilized in the autoclave.

Supplement

Horse colon liquor is essential in the medium for good growth of the isolates. It can be obtained by diluting colon contents 1 in 10 in normal saline, and straining through four thicknesses of muslin. The strained fluid can then be sterilized by passing through a Seitz filter. It is also stable to autoclaving at 121° for 15 min. It is added to the medium to give a final concentration of 1·0% (v/v).

Cellulose

This is most satisfactorily added to the medium as a finely divided suspension to give a final concentration of 0·1% (w/v). The suspension is prepared according to the method of Skinner (1960). Whatman ashless cellulose paper (standard grade) is rendered into a slurry by 20 min in a macerator in the proportion of 15·0 g in 80 ml distilled water. The slurry is divided into several aliquots and each shaken in 250 ml distilled water in a narrow container such as a measuring cylinder. When the deposit settles the supernatant is syphoned off and the deposit re-suspended in the same volume of water. This process is repeated until the supernatants appear only slightly turbid. The fluids are then bulked and made acid with a few drops of NHCl and allowed to stand overnight, when the cellulose flocculates. This flocculated deposit is concentrated by centrifugation and washed with several changes of distilled water until the deflocculation point is reached. This is indicated by a slight turbidity of the centrifugate. The deposit of cellulose is then re-suspended in distilled water to give a suspension of 2·0% (w/v). No sterilization is required. It is added to the medium to give a final cellulose concentration of 0·1% (w/v).

For concentrating cellulolytic bacteria in colon contents prior to isolation, strips of Whatman No. 1 filter paper are used. A strip c. 4 cm in width is wrapped round a glass rod of the length to fit into the test tubes to be used, and secured with nylon thread at both ends. Alternatively, very small pieces of paper can be used to fill the test tube to depth of about 1 cm. This paper is sterilized along with the test tubes in a hot-air oven at 160°. Cotton thread coiled into small hanks, *c.* 5 cm in diam, can be used to assess rate of cellulolysis in colon contents. These hanks are prepared by coiling and

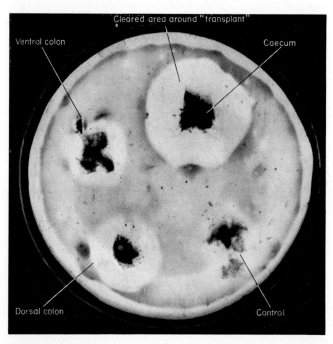

FIG. 1. Paper fragments impregnated with cellulolytic bacteria from caecum and ventral and dorsal colons, surrounded by areas of clearing due to cellulolysis in the medium. Paper control incubated in salt solution shows no area of clearing.

securing with nylon thread, boiling in three changes of distilled water, drying to constant weight and sterilizing along with the test tubes.

Isolation Technique

The basic salt solution, supplement and source of cellulose may be prepared some time before an experiment. The base can be stored at room temperature, in bottles filled to capacity until it develops a pink colour or precipitation appears. This is usually after a period of several weeks. The supplement can be stored in the refrigerator for a period of three months while the cellulose preparations can be stored indefinitely.

On account of the strict anaerobic character of cellulolytic bacteria, however, the complete medium must be prepared as rapidly as possible immediately before use.

Cultures prepared directly from samples

For these tenfold dilutions of the colon contents in normal saline are made in universal containers. These are clamped to a frame and linked in series by rubber tubing connected to large syringe needles passed through perforations in the bottle caps, to allow the flow of O_2-free CO_2 through the system. As each dilution is prepared a sample is withdrawn to inoculate tubes of culture medium, 0·2 ml of dilution being added to 1·8 ml melted, cooled medium in a test tube being flushed with O_2-free CO_2. The tube is then stoppered with a rubber bung fitted with a Bunsen valve, and the inoculated media allowed to set in the form of a roll-tube. Incubation is done at 39°.

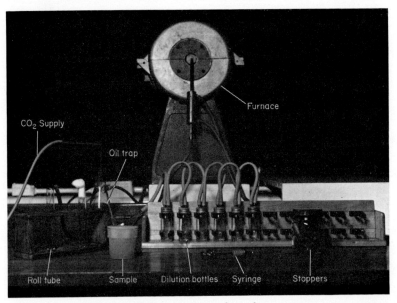

FIG. 2. Dilution apparatus for colon contents.

Concentration of cellulolytic bacteria in colon contents prior to culture

This has proved the most satisfactory of the isolation techniques used so far. It is modified from that of Khouvine (1923).

Test tubes containing filter paper on glass rods, as already described, are filled to capacity with colon contents diluted 1:10 in normal saline and stoppered with rubber bungs fitted with a Bunsen valve, the procedure being carried out under O_2-free CO_2.

When the paper shows signs of disintegration, usually after 7–10 days of incubation at 39°, it is removed from the glass rod and washed three times in distilled water to free it from extraneous matter. Fragments are then

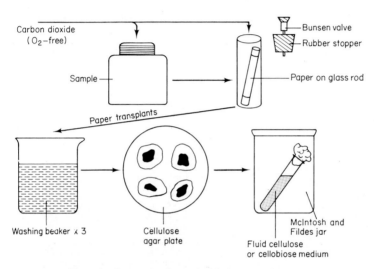

FIG. 3. Concentration of cellulolytic bacteria.

implanted by means of fine forceps on the surface of Hungate medium plates. These are incubated at 39° in McIntosh and Fildes jars in which 5% of the hydrogen has been replaced by CO_2.

After incubation periods varying from 10–20 days, areas of clearing can be seen in the cellulose medium around the paper fragments. Cellulolytic bacteria can be subcultured from these areas to Hungate medium in plates or roll-tubes, using an inoculating loop to scrape the surface of the cleared area.

Individual colonies which develop on the medium may be surrounded by a zone of clearing, but in many cases this is difficult to see. Alternatively, there may be a small hollow or depression in the medium around the colony.

For further study colonies may be transferred to fluid Hungate medium and subcultured alternately through fluid medium in which cellulose has been replaced by cellobiose in a final concentration of 0·1% (w/v).

To investigate the biochemical character of strains isolated, fluid Hungate medium can be used in which the cellulose is replaced by any one of a range of fermentable substances in a final concentration of 0·1% (w/v).

Precautions for strict anaerobiosis as already described are still required for these tests.

These methods have been used to isolate cellulolytic bacteria from the pig, rabbit and guineapig as well as from the horse.

Horse colon liquor has proved to be a satisfactory supplement for cultures from these species, and it may be replaced satisfactorily by rumen fluid as a supplement for all species (Davies, 1968).

It therefore seems probable that this technique could be used for the isolation of cellulolytic bacteria from other species, with the medium subject to little, if any, further modification.

The Artificial Colon

Another method of culturing bacteria from colon contents is available by use of the artificial colon. In the container vessel of this apparatus, a

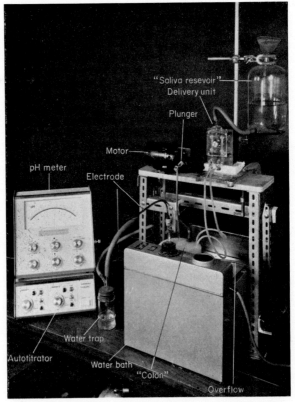

FIG. 4. Key to artificial colon.

continuous culture of a large volume of colon contents can be studied, and samples withdrawn for subculture on Hungate's cellulose medium for the isolation of individual organisms.

The main source of cellulose in the artificial colon is a good quality hay. Other sources such as paper and cotton thread can be substituted, and other substrates such as sugars and starch can be added to the system for the study of their influence on the composition of the bacterial population.

This method offers the advantage of culture on a larger scale than in the test tube, and where conditions approaching those believed to exist in the living host can be more nearly simulated, but it has also the disadvantage of a relatively complicated apparatus.

Acknowledgement

The Author has pleasure in thanking Mr R. K. Thompson of the Photographic Department, Royal (Dick) School of Veterinary Studies for taking the photographs.

References

DAVIES, M. E. (1964). Cellulolytic bacteria isolated from the large intestine of the horse. *J. appl. Bact.,* **27,** 373.

DAVIES, M. E. (1968). Role of colon liquor in the cultivation of cellulolytic bacteria from the large intestine of the horse. *J. appl. Bact.,* **31,** 286.

HUNGATE, R. E. (1950). The anaerobic mesophilic cellulolytic bacteria. *Bact. Rev.,* **14,** 1.

KHOUVINE, Y. (1923). Digestion de la cellulose par la flore intestinale de l' homme. *Annls. Inst. Pasteur, Paris,* **37,** 711.

SKINNER, F. A. (1960). The isolation of anaerobic cellulose-decomposing bacteria from the soil. *J. gen. Microbiol.,* **22,** 539 (also this book, p. 57).

The Isolation of Anaerobic Organisms from the Bovine Rumen

M. J. Latham and M. Elisabeth Sharpe

National Institute for Research in Dairying, University of Reading, England

The rumen supports a population of mixed organisms most of which are obligate anaerobes which will not grow under conventional anaerobic conditions. These bacteria are apparently unable to reduce a medium themselves (Hobson and Mann, 1970) and exist naturally in an environment where the Eh is kept at a low level by the metabolism of the accompanying bacteria and protozoa. In order to initiate growth both in their isolation and in pure culture they therefore need a highly reduced medium from which oxygen is excluded. In 1950 Hungate described fundamental techniques for the isolation of cellulolytic anaerobic rumen organisms. These were based on attempts to simulate *in vitro* the physical and chemical conditions found in the rumen, especially the highly reduced conditions. With some modifications (Bryant and Burkey, 1953; Bryant and Robinson, 1961) these methods have been used for the isolation and cultivation of all groups of strictly anaerobic bacteria found in the rumen. Essentially they comprise the final preparation of media in an anaerobic atmosphere and without further exposure to oxygen, an Eh poised at a suitable value by reducing compounds, anaerobic sampling and dilution of samples, and the use of media of composition similar to that of the rumen contents.

Materials and Equipment

Oxygen-free gases

Commercial gases, even those of high purity, frequently contain amounts of oxygen sufficient to prevent reduced conditions being maintained in media to which they are exposed.

Two methods can be conveniently used for preparing oxygen-free gases:

(a) Contaminating oxygen can be removed by combining with heated copper turnings to form copper oxide. The gas to be purified is passed slowly through a vertical glass column containing copper turnings heated to between 300–350° (Fig. 1). The column consists of an inner glass tube (70 × 2·8 cm) filled with copper turnings (BDH), which are supported at the lower end by a tight roll of brass gauze. Four strands of thin asbestos

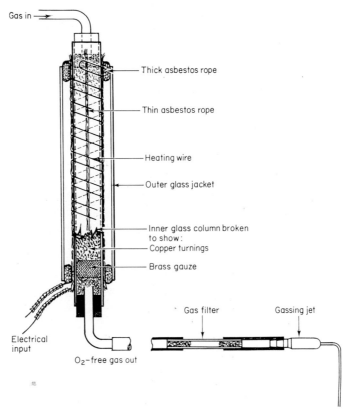

Fig. 1. Column for producing oxygen-free gas.

rope are strapped down the outside of the column and a heating element of 0·022 in. diam nickel-chromium wire (Brightray, Henry Wiggin Alloys) is wound tightly around the column at approximately ½ in. spacing to the height of copper turnings, the coils of heating wire being held apart by the asbestos rope. Thick asbestos rope wrapped around each end of the inner column locates it centrally within an outer protective glass jacket (50 × 4·0 cm). The exposed ends of the heating wire are covered in fish spine porcelain beads and connected via porcelain connectors to a variable transformer (Variac (8 amp), Claude Lyons Ltd.) which regulates the column temperature. During warm-up of the apparatus (15 min) gas is flushed through the column to remove any air which might have entered the system, during shut down.

The operating flow rate for gas should not exceed 400 ml/min as at higher flow rates the removal of oxygen is incomplete. As oxygen is removed from the gas the oxide formed causes the packing to darken progressively until its reducing capacity is exhausted. When a commercial "oxygen-free" gas is used the columns will last for about 8 h of continuous use before requiring regeneration. The hot column is regenerated by flushing with $90\% \; H_2 + 10\% \; CO_2$ which, for safety, must be vented to the external atmosphere. Reduction of the copper to a salmon pink colour is normally complete within a few minutes.

(b) This method depends on the catalytic combination of any contaminating O_2 with H_2 to form water. A special gas mixture of $99\% \; CO_2 + 1\%$ H_2 (British Oxygen Company) is passed through a cartridge containing paladinized asbestos (Deoxo Catalytic Purifier, Engelhard Industries Ltd). Any oxygen contained in the gas mixture is removed and the effluent gas used for cultural work. In this combination CO_2 and H_2 separate if markedly compressed, consequently cylinders cannot be filled to high pressures and thus empty rapidly. It would therefore be preferable to meter accurately into the cartridge the correct amount of each gas from separate cylinders. Provided the catalysts are not poisoned they are long lasting and can accept high flow rates without loss of efficiency.

The O_2-free gas from either source is distributed over short distances of up to 10 feet through rubber tubing. For longer distances copper tubing would be preferable. Sterile glass tubes containing cotton wool filters are inserted into the ends of each gas line and gassing jets are then attached.

Gassing jets

Short jets can be made by bending 3 in. No. 19 hypodermic needles through 90° close to the butt. More durable jets are constructed of 5 cm. and 15 cm lengths of narrow bore (1·5 mm i.d.) stainless steel tubing braised to a tapered stainless steel butt (4·3 cm long). A short length of rubber tubing connects the butt of the jets to the gas filters. All the jets with their rubber tubing connectors are stored in 95% ethanol when not in use.

Gas mixtures

Apart from the special gas mixture used for the catalytic removal of oxygen, the most frequently used gases are CO_2, 100%; N_2, $90\% + CO_2$, 10%, and N_2, 100%. N_2 or N_2/CO_2 gases are used where media of low buffering capacity are required and in certain *in vitro* incubations in which CO_2 would be inhibitory, e.g. the determination of microbial hydrogenating activity. For all routine cultural work however CO_2 is used. H_2/CO_2 is employed in the cultivation of the methanogenic bacteria (Smith and Hungate, 1958).

Media containers

The test tube sizes used are: agar slopes ($\frac{5}{8} \times 4$ in.), broths ($\frac{5}{8} \times 6$ in.) and roll-tube cultures ($\frac{3}{4} \times 6$ in.). Glass bottles (4 oz) are used for 100 ml quantities of media. Rubber bungs No. 10A (Esco) No. 4 and 17 (Baird and Tatlock) are used for closure of the tubes and bottles and are sterilized in long boiling tubes.

Roll-tubes

The roll-tubes require a wide orifice to accommodate the gassing jet and at the same time allow colonies to be picked. For rolling the $\frac{3}{4}$ in. tubes a simple apparatus has been constructed (Fig. 2), comprising an electrically driven gang of 6 horizontal detachable heads each with 4 spring "fingers" of stainless steel rod which grip the tubes from the base. Cold water is sprayed vertically downwards from 6 overhead pierced stainless steel tubes.

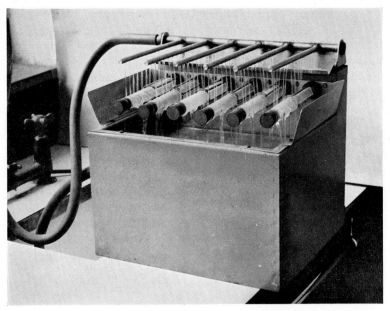

FIG. 2. The roll-tube apparatus.

Reducing agents

A mixture of cysteine-HCl and sodium sulphide (0·015% w/v, each final concentration) is used routinely as the reducing agent for most media with the addition, mainly to roll-tube media, of dithiothreitol (0·0001 M, final concentration), a protective reagent for SH groups (Cleland, 1964), to help poise the media at a low Eh. Cysteine-HCl alone (0·05% w/v, final concentration) is used in media for the detection of sulphide production.

Reduction indicator

Resazurin (0·0001% w/v) is incorporated into media as a redox indicator. Reduction takes place in two stages; resazurin (blue)⟶ resorufin (pink) ⇌ dihydroresorufin (colourless). The first stage is irreversible whereas the second stage is readily reversed by traces of oxygen. However, it is not an ideal indicator of the state of reduction since the oxidation–reduction reaction of resorufin to dihydroresorufin has the relatively high calculated E_0^1 of $-0·042$ volts at pH 6·87 (Twigg, 1945). Other indicators which may reduce at lower potentials are reported to be toxic unless diluted too much to observe their colour changes. Phenosafranine is preferred in studies on the methanogenic bacteria (Bryant, 1963).

Methods

The methods described here are known as the open-tube technique i.e. all inoculation and transfers, although still performed anaerobically, take place with the rubber bung removed. A method with less inherent danger of oxidation or contamination is the closed-tube technique in which the bung is never removed and inoculation is performed via injection with a hypodermic syringe through a recessed bung. For routine laboratory use, especially when dealing with large numbers of cultures, the open-tube technique is more practicable.

Preparation of media

Formulae of essential media as used in this laboratory for the isolation, viable counts and cultivation of rumen bacteria are given in the Appendix. They consist of: (i) a dilution fluid for the serial dilution of the rumen sample, and (ii) a general medium for the isolation and cultivation of most types of rumen organisms. This may be a medium containing rumen fluid (Bryant and Robinson, 1961) or one in which the rumen fluid is replaced by a number of volatile fatty acids and haemin (Caldwell and Bryant, 1966).

The latter medium has the advantage that it does not have the variation in composition which may arise from the use of different samples of rumen fluid. Colonies are picked from the general medium into agar slopes of the same composition. Other media used for identification of the isolates are based on the general medium chosen, with relevant additions or omissions of ingredients. All media are reduced by a combination of reducing agents

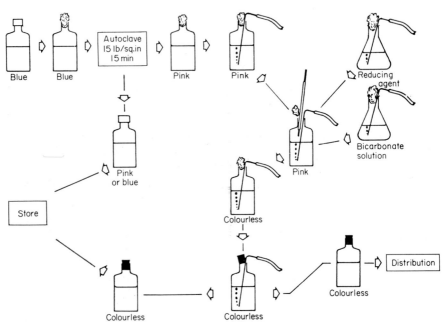

FIG. 3. Flow diagram for the reduction of media showing colour changes of the redox indicator (resazurin).

and oxygen-free gas, reduced conditions are maintained by an oxygen-free atmosphere and the state of reduction is made apparent by the colour reaction of a redox potential indicator.

Whenever possible media are prepared in bulk and distributed in the reduced state. Media correctly prepared will keep at room temperature for several months. A flow diagram (Fig. 3) for the reduction of media is given.

All ingredients, except agar and those to be added after sterilization are combined together and adjusted to pH 6·8–7·0 with NNaOH. Buffering capacity is usually provided by sodium bicarbonate (0·5 % w/v) equilibrated with CO_2 to give a pH of c. 7·0. However, if media of low buffering capacity are required the bicarbonate is reduced to 0·06 % (w/v) and the gas phase changed to nitrogen. The amount of KH_2PO_4 and K_2HPO_4 present in the mineral solutions provide some slight buffering capacity. After adjusting the pH the final volume is made up with water. Allowance must be made for the addition after autoclaving of bicarbonate and reducing agent solutions, and in some instances sugar solutions or other additives e.g. triglyceride emulsions. The medium is distributed into 100 ml screw-capped bottles or into larger flasks if large quantities of bacterial cells in batch or continuous culture are required. Agar is distributed to the bottles as required before the addition of medium. The filled bottles are usually sterilized with cotton-wool plugs but if the medium is to be stored, loosened screw caps are used. After autoclaving and without being reduced, the bottles can then be allowed to cool to room temperature, tightly sealed, and stored. When required, the media are steamed, gassed and reduced in the normal fashion. For immediate use freshly autoclaved media are allowed to cool a little and equilibrated with O_2-free gas with the cotton-wool plugs still in place. For broth media, long gassing jets penetrating deep into the medium are used, while agar media can only be gassed at the surface by short jets. The reducing agent and bicarbonate solutions sterilized by filtration are also gassed thoroughly before being added by pipette to the hot media. The pipette is held with its tip in the gas stream just above the surface of the sterile solution and all the air within the pipette is sucked out until the CO_2 can be tasted. If nitrogen is being used several volumes are sucked through the pipette. The pipette is then filled with the appropriate volume of liquid, quickly transferred to the hot medium and the contents are allowed to run out but are not blown out. Gassing is continued in broth media for some minutes after the resazurin has turned colourless. The cotton wool plug is replaced by a sterile rubber bung which is inserted into the neck of the flask as the gassing jet is slowly removed. Agar media, since they are only being gassed at the surface, are sealed immediately after the addition of reducing agents, bicarbonate etc, inverted several times and placed in a 45–50° water bath. Reduction of resazurin is complete within a few minutes.

Distribution of media

This is perhaps the most difficult part of the whole anaerobic procedure and our limited attempts to produce a satisfactory automatic medium dispenser have failed, due to excessive contamination with air. Although dispensing media by hand is very tedious, it is still in our experience the only way to maintain the best reduced conditions.

Sterile tubes are fitted with sterile rubber bungs, and filled from a bottle of the reduced medium. A gassing jet is inserted into the bottle and another into each tube as they are filled out. To do this a test tube is taken in one hand and the bung removed between finger and thumb whilst holding the tube firmly against the palm of the hand with the other fingers. A gassing jet is hooked over the lip of the tube with the other hand and left in position so that the tube is flushed with a continuous stream of gas. Reduced medium is drawn up into a pipette using the precautions previously described and transferred to the tube. The pipette is discarded, the tube gassed for a further 10–15 sec and the bung re-inserted as the gassing jet is withdrawn.

Isolation of strains

Samples of well-mixed rumen contents are taken by hand from fistulated animals to fill completely a wide-necked screw-capped jar or polythene bottle. Rumen fluid for media preparation is collected free of large particles by suction through a nylon mesh filter suspended within the rumen in a weighted stainless steel cage. If anaerobiosis is to be maintained in collected rumen material during transfer to the laboratory only a very small head space is left in the collecting vessel and O_2-free gas bubbled in continuously from a football bladder inflated with the O_2-free gas. A completely filled sealed jar will maintain its own anaerobic state satisfactorily. From the time the rumen sample is removed from the animal all processes must be carried out under strictly anaerobic conditions.

A 10 g core of rumen ingesta is taken from the centre of the sample jar and dispersed in 90 ml of anaerobic dilution fluid (Bryant and Burkey, 1953) by high-speed mixing in a MSE homogenizer, the whole process being kept anaerobic by constant gassing with CO_2. Serial dilutions are made in further quantities of the dilution fluid. The dilution fluid is the most susceptible to oxidation of all the media employed, probably due to the absence of reducing sugars.

The reduced medium for roll-tubes previously distributed in 9 ml quantities and being held in a water bath at 45–50° is inoculated with 1 ml of the required dilution. Again, very thorough gassing is important as ultimately a large surface area of agar will be exposed to the gaseous atmosphere within the tube and any residual oxygen will cause the medium to oxidize. The tubes are rolled as soon as possible after inoculation as prolonged holding in the water bath results in a reduced viable count. A more uniform agar layer and a better seal are obtained if the tubes are inverted several times before rolling. Once rolled the tubes are incubated vertically. On no account should the tubes be tipped on their side during or after incubation as the condensation which forms within the tube will wash over surface colonies so contaminating the entire tube. A typical roll-tube prepared from diluted bovine rumen contents together with an

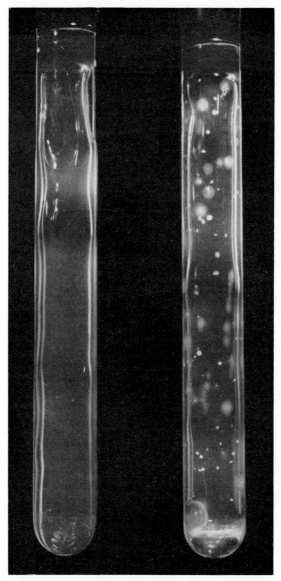

F IG. 4. Two Medium 10 roll-tubes: uninoculated tube showing gas pockets (left), inoculated tube showing many different colony types (right).

uninoculated tube are shown in Fig. 4. Some spreading colonies are visible and in the uninoculated tube a particularly bad example of gas pocket formation can be seen.

Picking colonies

Colonies are picked from tubes clamped against a dark background and illuminated by a diffuse light source. After swabbing the outer surface of the bung with alcohol and brief flaming, the bung is removed slowly since a sudden inrush of air will oxidize the agar and may also cause the agar to implode. A short gassing jet is then clamped within the tube but not touching the agar. The thicker agar at the top of the roll-tube hinders the introduction of the picking needle and soon becomes charred as a result of the frequent flaming of the mouth of the tube. Consequently, before picking, the top half inch of agar is removed. Once the tube has been opened, the colonies must be picked immediately as prolonged exposure to the gas stream causes the agar to dry. For picking colonies, an agar slope is taken in one hand, the bung removed between finger and thumb and a short gassing jet inserted. The bung whilst still being held is allowed to rest in the tube orifice and against the gassing jet whilst a colony is picked. A platinum needle with the tip bent at right angles is used for picking and the growth quickly transferred to the slope, stabbing deep into the butt of agar. The slope is then gassed for a further minute and the bung replaced as the jet is removed.

Transfer of organisms

Organisms which are killed by oxygen as opposed to being merely inhibited must be transferred using well-tapered Pasteur pipettes fitted with rubber teats. Air is replaced by O_2-free gas by repeatedly squeezing the bulb with the tip held close to the end of the gassing jet. A small volume of culture broth, water of syneresis or a plug of agar, is drawn up followed by a further quantity of O_2-free gas so sandwiching the growth in an anaerobic atmosphere. Transfer can then be completed with very little risk of oxidation.

Storage of cultures

Since most pure isolates of rumen organisms will not remain viable at $4°$ or freeze dry successfully, they have to be kept deeply frozen at $—65°$ in an insulated box containing solid CO_2. However, some strains of *Bacteroides succinogenes*, *B. amylophilus* and *B. ruminicola* have recently been successfully freeze dried in this laboratory but as yet this technique has not been used with the majority of rumen organisms.

Starting up an organism from the frozen state

This is best performed as soon as the cultures have thawed to room temperature, transferring growth with either a platinum loop or Pasteur pipette. Nickel-chromium wire, particularly when pitted, should not be used, since it can cause slight oxidation of broth media.

Precautions

The technique frequently poses difficulties when first attempted but with strict attention to details success can quickly be achieved. Most problems are caused by the inability to obtain sufficiently reduced media or by reoxidation of media after distribution.

Reduction of media

The greater the quantity of medium the more it needs to be gassed. During routine preparation it is preferable to prepare sterilized non-reduced media in bulk and distribute only small quantities of reduced media at any one time. Since every tube requires individual attention to the gassing, the distribution of media is the most exacting part of the technique. Gassing should always be continued for some time after the reduction of the resazurin, depending on the volume of medium and its head space and should continue throughout the distribution procedure. Failure to replace the air in the mouth pipettes with O_2-free gas will result in that aliquot of medium oxidizing within the pipette. Likewise if the medium is blown from the pipette oxidation will occur within the tube. Whenever possible a fresh pipette should be used for every new aliquot of medium as some residual medium is certain to be left in the tip of the pipette and become oxidized in the brief period of exposure to air.

Reoxidation

Oxidation of distributed media may be the result of either poorly sealing bungs or undetected reduction in the gas flow during distribution. Any bungs with scored surfaces or which have become badly discoloured as a result of autoclaving should be replaced, even a small flaw will admit air. Likewise no tube with a chipped rim should be used as this will cut into the bung as it is twisted in. Full cylinders of CO_2 are prone to form ice in the regulator if the flow rate is kept too high. Thus, with too little gas entering a tube appreciable quantities of air may remain. The oxidation of media has also been attributed to the use of red rubber bungs (Hungate, 1963) which are more permeable to air. Butyl rubber bungs are therefore recommended. However, we have found red rubber bungs (Baird and Tatlock, No. 4 and 17) satisfactory, but for closure of $\frac{5}{8}$ in. test tubes rubber of a different composition (Esco, No. 10A) is preferable.

Gas-producing organisms blow out their bungs with considerable force not only admitting oxygen but also creating an aerosol. To prevent this, the bungs, particularly in broth media containing high sugar concentrations, are taped on with masking tape. Bungs are also taped on before storage in solid CO_2 as they tend to lose their elasticity and can easily drop out at the low temperature.

Contamination

After distribution all media are incubated at 37° for 24 h and any tubes showing oxidation or growth are discarded. No contamination has resulted from the gas system, but contamination with micrococci may sometimes occur from handling the rubber bungs. Contaminated isolates are purified by re-isolation from roll-tubes or, in the case of the bacteroides contaminated with micrococci, by growth in broth containing Vancomycin (5 μg/ml). In practice the incidence of contamination is very low.

Picking colonies from roll-tubes

A common fault with roll-tubes is the formation of pockets of gas between the glass and the agar towards the top of the tube (see Fig. 4) which increase considerably the danger of an implosion of agar when the bung is removed. In addition, rapidly spreading colonies which can encircle the tube (lachnospira), run down the agar (*Streptococcus bovis*) or spread as a very thin film (selenomonads) may contaminate or prevent the growth of many other colonies. Thus all roll-tubes must be examined carefully before making any isolations. Growth of the spreading organisms may be retarded by restricting the total carbohydrate concentration in the medium to 0·2% (w/v) or less.

Conclusions

The success of the Hungate technique lies in its basic simplicity. The possibility of exploring a number of anaerobic habitats where previous methods have failed is ample justification for the increasing acceptance of the simple but laborious technique. Our own work involving the characterization of the rumen flora of cows receiving many different rations has meant that our methods have, because of the very large quantities of media required, departed slightly from the original (Hungate, 1950) and subsequently modified methods (Bryant and Burkey, 1953). Whilst the methods described here are satisfactory for the isolation of most groups of anaerobic rumen organisms, culturing the even more strictly anaerobic methanogenic bacteria would require additional modifications to produce more highly reduced conditions.

Appendix

Media, mixtures and solutions

Mineral I

K_2HPO_4	0·6%	w/v in distilled water

Mineral II

NaCl	1·2%	
$(NH_4)_2SO_4$	1·2%	
KH_2PO_4	0·6%	w/v in distilled water
$CaCl_2$ anhydrous	0·12%	
$MgSO_4.7H_2O$	0·25%	

Volatile fatty acid mixture (VFA)

Acetic acid	17 ml
Propionic acid	6 ml
n-Butyric acid	4 ml
iso-Butyric acid	1 ml
n-Valeric acid	1 ml
iso-Valeric acid	1 ml
DL-α-methyl butyric acid	1 ml

*Cysteine—Na_2S reducing mixture

Cysteine HCl	1·5%	
$Na_2S.9 H_2O$	1·5%	w/v in distilled water

Dissolve the cysteine in a little distilled water, adjust pH to 10·0, using 10NNaOH, add sulphide, make up to volume, Seitz filter.

Haemin

1% (w/v) solution in 50% (v/v) ethanol + 50% (v/v) 0·05 N NaOH.

Resazurin

0·1% (w/v) in distilled water.

---

* Make up fresh each time a medium is prepared. The other solutions can be stored at 4° and used as required.

*Cysteine

2·5% (w/v) in distilled water, Seitz filter.
*NaHCO$_3$ or Na$_2$CO$_3$
8% (w/v) in distilled water, Seitz filter, or autoclave (121°/15 min).

Dilution Solution

Mineral I	7·5%	
Mineral II	7·5%	v/v
Resazurin solution	0·1%	

Make up to volume less 7% with distilled water, autoclave (120°/15 min), gas. Add cysteine solution (2%, v/v) and NaHCO$_3$ solution (5%, v/v) whilst gassing. Continue gassing until reduced and seal.

Media for roll-tubes

(a) Rumen fluid medium

Glucose	0·05%	
Cellobiose	0·05%	w/v
Soluble starch	0·05%	
Mineral I	7·5%	
Mineral II	7·5%	
Rumen fluid	20·0%	v/v
Resazurin solution	0·1%	

Dissolve ingredients and adjust to pH 6·8.
Add 2·0% (w/v) agar. Make up to volume less 7% with distilled water. Autoclave (121°/15 min), gas.
Add cysteine-Na$_2$S (2%, v/v) Na$_2$CO$_3$ (5%, v/v) whilst gassing. Continue gassing and seal.

(b) Medium 10 (M10)

Glucose	0·05%	
Cellobiose	0·05%	
Starch	0·05%	w/v
Yeast extract	0·05%	
Trypticase	0·2%	
Resazurin solution	0·1%	
Haemin solution	1·0%	
VFA mixture	0·31%	v/v
Mineral I	7·5%	
Mineral II	7·5%	

* Make up fresh each time a medium is prepared. The other solutions can be stored at 4° and used as required.

Dissolve ingredients and adjust to pH 6·8. Add 2·0% (w/v) agar. Make up to volume less 7%, with distilled water.

Autoclave (121°/15 min), gas.

Add cysteine-Na$_2$S. (2%, v/v) Na$_2$CO$_3$ (5%, v/v) whilst gassing. Continue gassing and seal.

M10 medium for slopes

As for (b) above but with

Glucose	0·1%	
Cellobiose	0·1%	
Starch	0·1%	w/v
Maltose	0·1%	
and Agar	1·5%	

References

BRYANT, M. P. (1963). Symposium on microbial digestion in ruminants: Identification of groups of anaerobic bacteria active in the rumen. *J. Anim. Sci.*, **22**, 801.

BRYANT, M. P. & BURKEY, L. A. (1953). Cultural methods and some characteristics of some of the more numerous groups of bacteria in the bovine rumen. *J. Dairy Sci.*, **36**, 205.

BRYANT, M. P. & ROBINSON, I. M. (1961). An improved non-selective culture medium for ruminal bacteria and its use in determining diurnal variation in numbers of bacteria in the rumen. *J. Dairy Sci.*, **44**, 1446.

CALDWELL, D. R. & BRYANT, M. P. (1966). Medium without rumen fluid for nonselective enumeration and isolation of rumen bacteria. *Appl. Microbiol.*, **14**, 794.

CLELAND, W. W. (1964). Dithiothreitol. A new protective reagent for SH groups. *Biochemistry*, **3**, 480.

HOBSON, P. M., & MANN, S. O. (1970). Special techniques for handling anaerobic bacteria. *J. Gen. Microbiol.*, **60**, 5.

HUNGATE, R. E., (1950). The anaerobic mesophillic cellulolytic bacteria. *Bact. Rev.*, **14**, 1.

HUNGATE, R. E. (1963). Polysaccharide storage and growth efficiency in *Ruminococcus albus. J. Bact.*, **86**, 848.

SMITH, P. H. & HUNGATE, R. E. (1958). Isolation and characterization of *Methanobacterium ruminantium* n.sp. *J. Bact.*, **75**, 713.

TWIGG, R. S. (1945). Oxidation reduction aspects of resazurin. *Nature, Lond.*, **155**, 401.

M

Isolation of Cellulolytic and Lipolytic Organisms from the Rumen

P. N. Hobson and S. O. Mann

Rowett Research Institute, Bucksburn, Aberdeen, Scotland

In a number of habitats occur anaerobic bacteria which cannot be isolated using conventional anaerobic techniques. They may, however, be isolated using tube-culture techniques in which all preparation of media and all manipulations concerned with inoculation and transfer of cultures are carried out under a stream of oxygen-free gas. This technique has been mostly applied to culturing of rumen bacteria, but it is finding greater use in examination of other microbial habitats. The methods to be described here utilize this technique. Some work suggests that similar degrees of anaerobiosis in plate cultures may be obtained by use of anaerobic box techniques (see Draser, p. 93 this volume), and so the solid media described here might be used in plates, or the liquid media in tubes, manipulated in an anaerobic box.

Culture of cellulolytic and lipolytic bacteria

Anaerobic cellulolytic bacteria are important in digestion of plant fibres in the rumen and numbers and types need to be determined in many surveys of the rumen microflora. There has been controversy over the "cellulose" substrate needed to demonstrate true cellulolysis (i.e. the degradation of native plant fibres). The methods described here use prepared "cellulose", but although this may lead to some overestimation of the numbers of "cellulolytic" bacteria, at least all true cellulolytic bacteria will be included amongst those growing.

Practically all ruminant feeds contain glycerides of long-chain fatty acids which are hydrolysed in the rumen and lipolytic bacteria may be an important constituent of the microflora. The methods described here were designed to isolate and enumerate truly lipolytic bacteria; that is, those capable of hydrolysis of long-chain emulsified glycerides. Esterase activity could be determined by substituting triacetin or tributyrin for the substrates described.

The general methods and media might be used for isolation of cellulolytic and lipolytic bacteria from other habitats, but the media might have to be altered to contain sources of growth factors other than rumen fluid; for instance faecal extract for intestinal bacteria.

Although these media are not completely selective for cellulolytic or lipolytic bacteria, they are designed to produce a measurable change which indicates the presence of bacteria with these properties in growth of a mixed population, or which enables bacteria with these properties to be differentiated from others.

The use of these media in the investigation of rumen function has been described in more detail elsewhere (Hobson and Mann, 1961; Mann, 1968; Kurihara, Eadie, Hobson and Mann, 1968).

Methods

General

The techniques for preparing media and manipulating cultures under oxygen-free carbon dioxide are based on those originated by Hungate some years ago and most recently described by him in review articles (1966, 1969). Some modifications to the methods as described have been made by ourselves and other workers and the general techniques are also described in this volume (Sharpe and Latham, p. 133). Because of the long time of incubation of cellulose-containing cultures care must be taken to ensure that no air diffuses into the tubes through badly-fitting bungs or ones made from unsuitable rubber. Secure taping (see below) will ensure that bungs are not loosened by development of internal gas pressure.

There are three methods of determination of cellulolytic activity. One, the roll-tube method, is primarily used for counting and isolation of cellulolytic bacteria or purification of cellulolytic cultures. The filter paper-strip method can be used for counting cellulolytic bacteria, although the precision of the method will depend on the dilution steps used and it is often not possible to subculture pure strains from the primary cultures as, in theory, it is from roll-tube cultures. The method can also be used for qualitative testing of cellulolytic activity of pure cultures. The third method using a suspension of cellulose fragments, is primarily used for quantitative testing of cultures for cellulolytic activity.

Media and methods

Although some variations have been described the basal medium can be similar for all types of cellulose-containing cultures and only one is described in detail here. This contains a small concentration of cellobiose to initiate growth of the bacteria, although cellobiose is inhibitory to cellulolysis in high concentration. Failure to obtain cellulolytic activity could be caused by the cellobiose and for some bacteria the concentration of this may need to be further decreased, or it could be omitted. The nitrogen sources for growth include amino acids and ammonia, and the rumen fluid provides growth factors, such as branched-chain volatile fatty acids.

Basal medium for cellulolytic bacteria

The basal medium contains per 100 ml; Bacto Casitone, 0·25 g; Bacto yeast extract, 0·06 g; centrifuged rumen fluid, 10 ml; mineral solution A, 15 ml; mineral solution B, 15 ml; resazurin (0·1% (w/v) aqueous solution), 0·1 ml; cellobiose, 0·025 g; cysteine HCl, 0·05 g; sodium bicarbonate, 0·4 g; distilled water, 60 ml. Mineral solution A contains (g/litre): KH_2PO_4, 3; $(NH_4)_2SO_4$, 6; NaCl, 6; $MgSO_4$, 0·6; $CaCl_2$, 0·6. Mineral solution B contains (g/litre): K_2HPO_4, 3. Centrifuged rumen fluid is prepared by centrifuging rumen contents, strained through gauze, at about 20,000 rpm. for 15 min in a steam-driven Sharples Super Centrifuge.

The cellobiose, cysteine HCl and sodium bicarbonate are added in a concentrated filter-sterilized solution to the main bulk of the medium which is sterilized at 121° for 15 min. The medium is prepared under oxygen-free CO_2 (using techniques similar to those described by Sharpe and Latham; this volume, p. 133). Twelve millilitre amounts are tubed in 6 in. $\times$ $\frac{5}{8}$ in. test tubes, under CO_2, and closed with rubber stoppers.

Filter paper strip medium

Strips of Whatman No. 1 filter paper of an inverted T-shape and of the following dimensions are prepared. The upright leg of the T is $\frac{3}{8}$ in. $\times$ $3\frac{7}{16}$ in., and the cross-piece $\frac{7}{16}$ in. $\times$ $\frac{5}{16}$ in., with the width of the upright at the junction with the cross-piece narrowed down to $\frac{5}{16}$ in. Round the cross-piece is looped a thin nylon suture on which is threaded a $\frac{3}{4}$ in. length of glass capillary tubing. The other end of the paper strip is inserted $\frac{3}{16}$ in. into a slit cut across the base of a rubber bung, so that the whole can be inserted into a 6 in $\times$ $\frac{5}{8}$ in. test tube with the weight hanging freely and the bung lightly resting in the tube. The assembled tubes are sterilized at 121° for 15 min, and after removal from the autoclave the bungs are firmly pressed into the tubes.

Suitable amounts (1 ml) of dilutions of rumen contents or other material are added to the tubes of the basal medium and the bungs replaced by bungs carrying a filter paper strip. All operations are carried out under a stream of oxygen-free CO_2, and the bungs are taped in place to prevent "blow-out" by gas formation.

Cellulose roll-tubes

The medium is similar to the basal medium described above except that the concentrations of Casitone, yeast extract and rumen fluid are doubled and cellobiose is omitted. To the part of the basal medium sterilized by autoclaving is added agar (2 g/100 ml) and cellulose powder (1 g; MN cellulose 300, Camlab (Glass) Ltd., Cambridge). The medium is prepared as above and dispensed in 5 ml amounts in 6 in. $\times$ $\frac{5}{8}$ in. test tubes at 49°, the bulk medium being periodically shaken to prevent the cellulose from settling. Portions (0·5 ml) of diluted material are added to each tube, the tubes inverted once or twice and rolled in cold water, either by hand or mechanically.* The bungs are taped on and the tubes incubated in an upright position.

Filter paper powder medium

The "roll-tube" medium described above, but with agar omitted, can be used. For assessment by centrifuging (see below) it may be more convenient to prepare the medium and grow the cultures in small graduated centrifuge tubes.

Medium for lipolytic bacteria

The medium contains per 100 ml: mineral solution A, 15 ml; mineral solution B, 15 ml; centrifuged rumen fluid, 40 ml; water, 29 ml; resazurin (0·1% (w/v) aqueous solution), 0·1 ml; sodium bicarbonate, 0·4 g; cysteine HCl, 0·05 g (for solid media add agar, 2 g). The sodium bicarbonate and cysteine HCl are added in a concentrated filter-sterilized solution to the rest of the medium which is sterilized at 121° for 15 min. To this basal medium is added 2 ml of a 50% (v/v) emulsion of linseed oil in rumen fluid. The rumen fluid (5 ml) is sterilized in a small covered beaker at 121° for 15 min, cooled and an equal volume of unsterilized linseed oil added. The emulsion is prepared by shaking the mixture on a wrist-action shaker (Griffin and George, Alperton, Wembley, Middlesex) or better by ultra-sonic agitation. The latter is carried out by immersing the tip of a Soniprobe (Dawe Instruments, Concord Road, Western Avenue, London) to the oil-rumen fluid interface and agitating at full power for 30 sec. The Soniprobe is sterilized by standing in 95% (v/v) aqueous alcohol and washing with absolute alcohol.

The bulk medium is shaken to dispense the oil emulsion and dispensed into 6 in. $\times$ $\frac{5}{8}$ in. test tubes in 9 ml (for liquid) or 5 ml (for solid) volumes.

* See also p. 136.

Portions (1 or 0·5 ml for liquid or solid) of diluted rumen contents are added to the tubes of media, and after inversion two or three times roll-tubes are prepared from the agar-containing medium. Bungs are taped in place and the roll-tubes incubated upright. The liquid cultures are incubated horizontally on a slow rocking-shaker.

Unsterilized linseed oil is used to prevent changes which would occur on sterilizing. The oil is taken with a sterile pipette from about half way down the liquid in a large bottle kept at 4°. No contamination from the oil has been noted in cultures.

Assessment of Cultures

Cultures of rumen contents are all incubated at 38°, but, just as the medium constituents may need to be altered for bacteria from other habitats, so may the temperature and times of incubation.

Cellulolytic cultures

Paper-strip cultures

These are incubated for a maximum time of 21 days. Three tubes are inoculated from each dilution; for example, from 10^3 to 10^{10} of rumen contents. The highest dilution containing cellulolytic bacteria is that in which the filter paper strip is "pitted", or "pitted" and beginning to disintegrate (see Fig. 1). A dilution of rumen contents is only counted as positive if two or three of the triplicate cultures show cellulolytic activity.

Cellulose roll-tubes

These are incubated for a maximum time of 28 days. Cellulolytic colonies are those surrounded by zones of complete or partial clearing of the cellulose. This latter is often difficult to interpret and can lead to some discrepancies in counts between different observers.

Filter paper powder medium

These cultures can be incubated for times similar to those given above. The culture is then centrifuged for a standard time (determined by experiment) and the volume of residual paper powder is then compared with an uninoculated control. The paper may also be filtered off, washed, dried and weighed.

Lipolytic cultures

Roll-tube cultures

Very small colonies (visible through a lens), surrounded by small zones of clearing of the oil emulsion, are visible after 2 to 6 days' incubation (Fig. 2).

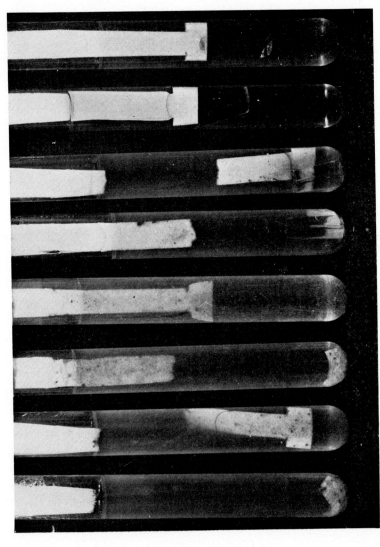

FIG. 1. A series of decimal dilutions (from left to right) of rumen contents showing disintegration and pitting of the filter paper. (Reproduced from *J. appl. Bact.*, **31**, 241, 1968.)

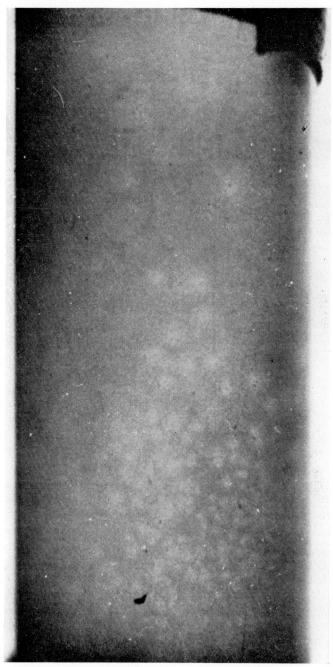

FIG. 2. Roll-tube culture of lipolytic bacteria in linseed-oil medium, showing zones of clearing of oil emulsion. (Approx. × 5 natural size.)

Liquid cultures

The tubes are incubated for 6 days. In this case either duplicate tubes of each dilution are prepared, one for testing lipolysis, one for sub-culture of the bacteria, or 1 ml amounts are removed as inoculum for later dilutions before testing the remainder of the cultures for lipolysis. The culture is brought to pH 5·6–5·8 by addition of HCl, extracted with ether, the ether extracts washed to pH 6·5–6·7 and the ether evaporated off. The long-chain fatty acid residue is dissolved in ethanol-water and titrated with standard NaOH.

Lipolytic activity of cell suspensions may also be tested by incubation in buffer with linseed oil, when the fatty acids formed can be determined titrimetically after extraction with isopropanol, heptane, sulphuric acid (Dole, 1956), or more rapidly by incubation with naphthyl stearate or laurate and determination of liberated β-naphthol with Fast Blue (Hobson and Summers, 1966).

Further Examination

Subcultures or Gram-stained films may be made from colonies showing zones of clearing of cellulose or linseed oil, although in the latter case the rumen lipolytic vibrios can be almost completely lysed or in spherical, granular form by the time that visible lypolysis has occurred. The latter forms will, however, regenerate on sub-culture.

Small portions of paper strips showing pitting may be removed aseptically and used as inocula for further cultures. However, as bacteria representative of those attached to the paper are free in the liquid, this may be used as inoculum for further dilution cultures to obtain pure strains.

Paper strips may also be stained by Gram's method with decolorization with acetone and counterstaining with safranin or dilute carbol fuchsin. The preparation is cleared with one or two drops of Whitemor 120 transparency oil (Shandon Scientific Co., London) and covered with a cover slip sealed with Vaseline.

A better method of staining to show bacteria attached to the filter paper is that of Baker and Nasr (1948). Two stains are required: No. 1 contains, 1% aqueous aniline blue, 2 ml; 5% aqueous phenol, 30 ml; glacial acetic acid, 8 ml; the mixture is allowed to stand for 1 h and then filtered, and No. 2 is a saturated solution of bismark brown in 95% aqueous alcohol. The staining procedure is: stain 1, 2 h at 37°; wash with 95% alcohol. 2 changes, for 15–30 sec; stain 2, 15–30 sec; wash with 95% alcohol, 15–30 sec, followed by absolute alcohol, 5 sec. The preparation is dried at 37°, cleared with xylol and mounted in Depex (G. T. Gurr, London). The

FIG. 3. Portion of filter paper strip from a culture, stained to show fibres with attached bacteria. (Staining method of Baker and Nasr, see text.) Magnification ×2700.

bacteria stain blue and cellulose structures yellow-brown (Fig. 3). The preparation is permanent and can be stored.

References

BAKER, F. & NASR, H. (1948). Microscopy in the investigation of starch and cellulose breakdown in the digestive tract, *J. R. Microsc. Soc.*, **67**, 27.

DOLE, V. P. (1956). A relation between non-esterified fatty acids in plasma and the metabolism of glucose, *J. clin. Invest.*, **35**, 150.

HOBSON, P. N. & MANN, S. O. (1961). The isolation of glycerol-fermenting and lipolytic bacteria from the rumen of the sheep, *J. gen. Microbiol.*, **25**, 227.

HOBSON, P. N. & SUMMERS, R. (1966). Effect of growth rate on the lipase activity of a rumen bacterium, *Nature, Lond.*, **209**, 736.

HUNGATE, R. E. (1966). *The rumen and its microbes*, New York and London: Academic Press.

HUNGATE, R. E. (1969). In *"Methods in Microbiology"*, Vol. 3B. London and New York: Academic Press.

KURIHARA, Y., EADIE, M. J., HOBSON, P. N. & MANN, S. O. (1968). Relationship between bacteria and ciliate protozoa in the sheep rumen, *J. gen. Microbiol.*, **51**, 267.

MANN, S. O. (1968). An improved method for determining cellulolytic activity in anaerobic bacteria, *J. appl. Bact.*, **31**, 241.

The Cultivation of Rumen Entodiniomorphid Protozoa

G. S. Coleman

*Biochemistry Department, Agricultural Research Council Institute of
Animal Physiology, Babraham, Cambridge, England*

Over the past 25 years rumen Entodiniomorphid protozoa have been
grown successfully *in vitro* several times. Hungate (1942, 1943) was the
first to grow these protozoa and he maintained several protozoal species for
over a year but only at low population densities. Gutierrez and Davis (1962)
cultivated *Epidinium ecaudatum in vitro* for 5 months but obtained populations of only 1200/ml. Clarke (1963) maintained *Eremoplastron bovis* for
5 months, while Mah (1964) maintained up to 1000 *Ophryoscolex purkynei*/
ml for nearly 3 years with a mean generation time of 24 h. The longest
period of continuous cultivation of a single species has been 10 years (1959
to the time of writing) for *Entodinium caudatum* (Coleman, 1958, 1960).
More recently *Entodinium simplex* has been cultivated for an unspecified
time by Jarvis and Hungate (1968) and for over $3\frac{1}{2}$ years (1966 to the time
of writing) by Coleman (1969a). The author has also maintained *Epidinium
ecaudatum caudatum, Polyplastron multivesiculatum, Ophryoscolex* sp. and a
diplodinium for shorter periods. This paper is an account of the author's
experiences in growing these rumen protozoa at Babraham rather than a
summary of the literature. Photomicrographs of some of these protozoa, all
at the same magnification are shown in Fig. 1.

Basic Conditions Required for Growth

Although the opinions of various workers in the field have differed as to the
necessity for certain additives to the growth medium, it is generally agreed
that these protozoa will only thrive under conditions that have been
designed to be similar to those in the rumen. As all the rumen protozoa are
strict anaerobes, care must be taken to exclude oxygen from the medium
for all but short periods. For protozoal survival a low redox potential must
also be established, and this is achieved by the addition of reducing agents
such as cysteine, by the growth of aerobic bacteria which are always present
in cultures as the result of chance contamination or by both agents together.

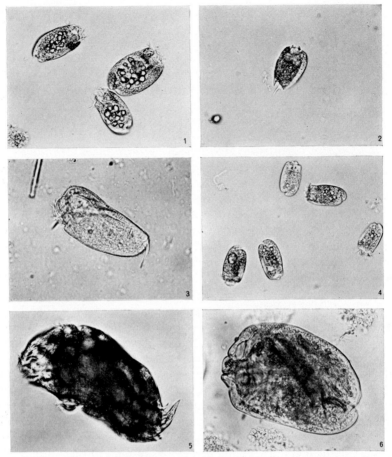

FIG. 1. Photomicrographs of rumen Entodiniomorphid protozoa cultured *in vitro*.
×247.

1. *Entodinium caudatum* after 10 years *in vitro*. 2. *Entodinium caudatum* taken directly from the rumen. Note the presence of the caudal spine which disappears after 1–2 years' cultivation *in vitro*. 3. *Epidinium ecaudatum caudatum* after 3 months *in vitro*. On further cultivation *in vitro* the caudal spine disappears. 4. *Entodinium simplex* after $3\frac{1}{2}$ years *in vitro*. 5. *Ophryoscolex* sp. 6. *Polyplastron multivesiculatum* grown for 3 months *in vitro* on Hungate-type medium.

It has proved impossible up to the present to grow these protozoa in the absence of bacteria and even in the absence of the specific addition of bacteria in e.g. fresh rumen fluid, there is always a thriving mixed bacterial population present with the protozoa. Because of the presence of bacteria in all cultures it is necessary to feed the protozoa with substrates that they can metabolize readily but which cannot be broken down readily by the bacteria. Therefore substances such as glucose which are poorly metabolized by Entodiniomorphid protozoa (Coleman, 1969b) but which stimulate bacterial growth cannot be used, whereas native intact starch grains, which are readily engulfed by the protozoa but are relatively resistant to bacterial degradation, are good substrates. Since heavy bacterial growth is always associated with death of the protozoa, it is necessary to maintain a balance between bacterial and protozoal growth by adding each day just sufficient starch grains for the protozoa to engulf without leaving an excess in the medium. The addition of twice the amount of starch every other day is not effective as the bacteria grow, even if only slowly, on the excess starch and the protozoa tend to die off.

Materials and Methods

Basal mineral salts solution

Hungate-type

This salts medium was based on that of Hungate (1942) and contained (g/100 ml): NaCl, 0·5; CH_3COONa, 0·15; K_2HPO_4, 0·1; KH_2PO_4, 0·03.

Coleman-type

This salts medium was based on that of Coleman (1958) and was usually made up at double the final concentration. The concentrated solution contained (g/100 ml): K_2HPO_4, 1·27; KH_2PO_4, 1·0; NaCl, 0·13; $CaCl_2$ (dried), 0·009; $MgSO_4$. $7H_2O$, 0·018. When this solution is prepared a precipitate will form unless care is taken not to add the calcium salt until the phosphates have been dissolved in the bulk of the water.

Rumen fluid for media

All rumen fluid for media was taken from Clun Forest sheep fed on 800 g hay and 200 g oats/day. The rumen fluid was taken 1 h after feeding and strained through one layer of muslin to remove the larger food particles. This strained fluid was centrifuged (500 **g**/5 min) to sediment the protozoa and the supernatant removed. If the rumen fluid was to be used fresh this supernatant was passed through a Ford's "Sterimat cc/o1" to remove any residual protozoa and stored under 95 % N_2 + 5 % CO_2 at 4° until required. Autoclaved rumen fluid was prepared by autoclaving (115° for 20 min) under 95 % N_2 + 5 % CO_2 the supernatant after centrifugation.

Dried Grass

Dried grass was prepared from lawn mowings. The material was initially air dried in thin layers on the bench and finally placed in an oven at 65° for up to 3 days. The dry grass was then finely ground in a Lee Attrition Mill (Lee Engineering Co., Milwaukee, Wisconsin, U.S.A.).

Starch

Rice starch (British Drug Houses Ltd., Poole, Dorset) or wholemeal flour (W. Prewett Ltd., Stone Flour Mills, Horsham, Sussex) was added as a 1·5% (w/v) suspension in water. The wholemeal flour suspension had to be added to the medium in single 1 or 2 ml quantities through a pipette with a large hole at the tip as the solid matter tended to sink rapidly to the bottom of the pipette if pipetting was carried out in the conventional way. As far as cultures of *Entodinium caudatum* were concerned, the use of rice starch which had been sterilized by heating the dry grains at 160° for 1 h resulted in the death of the protozoa. However the protozoa grew well in the presence of rice starch that had been heated dry at 120° for 24 h before use.

Gases

The author has always used $95\%N_2 + 5\%CO_2$ and "Medical quality" CO_2 as supplied by British Oxygen Co., without prior treatment. Removal of traces of oxygen by red-hot copper or chromous sulphate solution did not improve the growth of *Entodinium caudatum*.

Culture media

Caudatum-type

One litre flasks containing 250 ml concentrated Coleman salt solution, 250 ml H_2O and 5 ml 15% CH_3 COONa were autoclaved (115° for 20 min) and when cool the following additions were made: 8 ml 1·5% (w/v) rice starch, 2·5 ml 2% (w/v) L-cysteine hydrochloride (neutralized with NaOH immediately before use), 50 ml autoclaved rumen fluid and (for *Entodinium caudatum* only) 8 ml 1·5% (w/v) chloramphenicol (for the necessity of this compound, see *Entodinium caudatum*—initial isolation). The whole was then bubbled vigorously with $95\%N_2 + 5\%CO_2$ for 3 min before sealing with a rubber bung.

Simplex-type

One hundred ml centrifuge tubes containing 24 ml concentrated Coleman salt solution and 36 ml H_2O were autoclaved (115° for 20 min) and when cool the following additions made: 1 or 2 ml 1·5% (w/v) wholemeal flour, 10 ml autoclaved rumen fluid (or fresh rumen fluid where stated) 0·9 ml 2% (w/v) L-cysteine hydrochloride (neutralized with NaOH immediately before use) and 12 ml 5% (w/v) $NaHCO_3$ (freshly prepared in autoclaved water). The whole was then bubbled vigorously with CO_2 for 3 min before sealing with a rubber bung.

For some protozoal species e.g. *Epidinium ecaudatum caudatum* the salt concentration in this medium was too high and dilute Simplex-type medium in which 10 ml of the concentrated Coleman salt solution was replaced by water was used.

The use of these media in the routine maintenance of the protozoa is described under the individual protozoal species in the sections on "Daily Maintenance".

Protozoal counts

The number of protozoa was estimated by blowing 0·2 ml of the culture from a 1 ml graduated pipette with a large hole at the tip into 2·0 ml (for entodinia) or 0·2 ml (for epidinia) 0·02 M iodine and counting microscopically all the protozoa in 0·1 ml of the mixture. For *Polyplastron multivesiculatum* cultures where the numbers were usually less than 100/ml, 2 ml of the culture was pipetted into a 4 × ½ in. tube and all the protozoa present counted under an inverted microscope (Olympus Optical Co. Ltd., Tokyo—model CK). Only those protozoa which showed no signs of disintegration were counted. Unless otherwise stated the number of protozoa in a culture was always estimated immediately before dilution of that culture.

Measurement of redox potentials

Redox potentials were measured in completely filled and sealed tubes with an Ingold dual platinum electrode containing a Ag/AgCl half cell supplied by Pye Unicam Ltd., York Street, Cambridge. The results were corrected for the potential of +194 mv generated by the Ag/AgCl half cell.

Effect of the Various Constituents of the Media on Protozoal Growth

The detailed effect of varying in turn each material in the media on the growth of single protozoal species and of mixed cultures of protozoa is

N

shown below. It should be emphasized that all cultures contain a mixed bacterial population and that no precautions are taken to exclude bacteria.

Basal mineral salts solution

There are two types of basal mineral salts solution, the NaCl-rich one of Hungate (1942) and the potassium phosphate-rich one of Coleman (1958) and each favours the growth of different protozoal species. After inoculation with mixed Entodiniomorphid protozoa from the sheep rumen, the caudatum or simplex-type media which contain potassium phosphate-rich salts medium favoured the growth of entodinia whereas the NaCl-rich salts medium especially in the absence of rumen fluid and cysteine favoured the growth of *Polyplastron multivesiculatum* and other diplodinia.

The optimum salt concentration for protozoal growth depends on the partial pressure of CO_2 and this will be considered below.

Anaerobiosis

The rumen Entodiniomorphid protozoa are all strict anaerobes and incubation at 39° for 30 min in air-saturated salt solutions killed all the protozoa. It is essential that the redox potential should drop to 0 within 30 min of inoculation. Redox potentials greater than $+100$ mv are not compatible with the life of these protozoa. Fortunately most rumen protozoa are more tolerant of oxygen at room temperature and they can be washed or picked off a slide with a micromanipulator without loss of viability, provided that this is done as rapidly as possible.

Before the culture medium is inoculated it is necessary to remove free oxygen. This can be done by autoclaving the salt solution but for small volumes (i.e. up to 100 ml) bubbling nitrogen or other inert gas vigorously through the medium for 2–3 min is a better way of removing all the oxygen. For most protozoal species it is also advisable to add 0.01% (w/v) L-cysteine hydrochloride (neutralized) especially if only a small (less than 10%) inoculum from an existing culture is used. Although it is possible to maintain cultures by serial transfer of a 10% (v/v) inoculum once a week, the protozoa occasionally die for no obvious reason and it is safer to maintain established protozoal cultures by dilution with an equal volume of fresh medium twice a week. Under these conditions the bacteria in the existing medium ensure that the redox potential in the diluted culture is rapidly lowered to a level compatible with protozoal survival. To decrease the risk of exposure to oxygen most cultures are grown in almost completely filled centrifuge tubes and the air above the surface removed by passing in a vigorous stream of inert gas through a Pasteur pipette before pushing in a rubber bung as the pipette is removed. However there is some evidence that *Epidinium ecaudatum caudatum* grows more rapidly and to a higher population density in a 50 ml conical flask than in a 50 ml centrifuge tube. It is possible that this protozoon does not grow well when organisms are crowded together in a small volume at the bottom of a centrifuge tube.

Carbon dioxide

Entodinium caudatum grows equally well in caudatum-type medium, which is equilibrated with 95 %N_2 + 5 %CO_2 and in simplex-type medium which is equilibrated with 100% CO_2. In contrast *Entodinium simplex* grows well only in the latter medium and can only be grown with difficulty in the caudatum-type medium in the presence of 5% CO_2. *Polyplastron multivesiculatum* behaves curiously in that it grows in the absence of other protozoa in caudatum-type medium even when it contains no rumen fluid but will only grow in simplex-type medium in the presence of epidinia. The exact specificity of this requirement is unknown but the epidinia, which are engulfed by *Polyplastron multivesiculatum*, cannot be replaced by entodinia. *Polyplastron multivesiculatum* also grows well directly from the rumen in impure culture in Hungate-type salts medium (in the absence of rumen fluid or cysteine) equilibrated with 95% N_2 + 5% CO_2, but attempts to establish pure cultures with single organisms picked from this medium have failed. *Epidinium ecaudatum caudatum* from the sheep rumen grows in both the caudatum- and simplex-type medium but grows faster and to a higher population density in the latter e.g. 2300/ml from an inoculum of 100/ml in 10 days in the dilute simplex-type compared with only 960/ml in the caudatum-type medium. *Ophryoscolex* sp. grows better in the caudatum-type medium containing 10% (v/v) fresh rumen fluid.

The optimum salt concentration also varies with the partial pressure of carbon dioxide as was first shown for diplodinia by Hungate (1955). For *Epidinium ecaudatum caudatum* the optimum is 0·084 M (calculated from the composition of Coleman-type salts medium) in the presence of 5% CO_2 and 0·123 M (of which 0·077 M is $NaHCO_3$) in the presence of 100% CO_2.

Rumen fluid

Most Entodiniomorphid protozoa will grow in media in the absence of rumen fluid but the population density can be increased markedly except with *Polyplastron multivesiculatum*, by the addition of 10% rumen fluid. Fresh rumen fluid from which the protozoa have been removed is usually stimulatory to protozoal growth in the early stages after initial isolation but unfortunately can only be used immediately or after storage for only a few days under $95\% \, N_2 + 5\% \, CO_2$ at $4°$. In contrast, autoclaved rumen fluid which of course can be kept almost indefinitely, is not stimulatory until a culture has been maintained *in vitro* for some months. Within these generalizations different protozoal species behave in different ways with respect to the addition of fresh or autoclaved rumen fluid and the population densities attained by established cultures of four protozoal species are given in Table 1. For all the species there is a progressive increase in the

TABLE 1. Effect of rumen fluid on the growth of different species of Entodinio-morphid protozoa

Protozoon	No. of protozoa/ml medium* containing		
	No rumen fluid	10% fresh rumen fluid	10% autoclaved rumen fluid
Entodinium caudatum	2000	28000	26000
Entodinium simplex	3700		29400
Epidinium ecaudatum caudatum	1400	2140	2700
Polyplastron multivesiculatum	68	48	50

* The cultures of *Entodinium caudatum* and *E. simplex* were maintained in caudatum- and simplex-type media respectively by dilution with an equal volume of fresh medium every 2–3 days (Coleman 1960, 1969a) and the numbers estimated directly before dilution of the cultures. *Epidinium ecaudatum caudatum* was grown in dilute simplex-type medium from an inoculum of 30 organisms/ml for 10 days. *Polyplastron multivesiculatum* was grown in caudatum-type medium by dilution twice each week for two months with the daily addition of 10 epidinia/ml and the numbers estimated directly before dilution of the culture. The *Entodinium caudatum* cultures were fed with 0.12 mg rice starch/ml and the other protozoa with 0.25 mg wholemeal flour/ml daily.

protozoal growth rate as the concentration of rumen fluid is increased from 0 to 10%; any further increase in concentration has no effect or is inhibitory.

To determine whether the active material in the rumen fluid was soluble, the crude material was first centrifuged ($10,000 \, \mathbf{g}/20$ min). The pellet was then washed once in salt solution and resuspended in the original volume and both the original supernatant and the washed pellet autoclaved ($115°$ for 20 min) in sealed bottles under $95\% \, N_2 + 5\% \, CO_2$. When used in place of autoclaved rumen fluid in the normal growth media both fractions were stimulatory to the growth of *Entodinium caudatum* and *E. simplex* and for the latter the pellet fraction completely replaced the autoclaved whole rumen fluid. Although the present author has never investigated the effect of using rumen fluid from sheep on different rations and has obtained satisfactory results with rumen fluid from sheep fed on hay and oats, it should be noted that Tompkin, Purser and Weiser (1966) found that for the growth of entodinia, rumen fluid from sheep containing high population densities of entodinia was superior to that from sheep containing few entodinia.

Source of starch

The nature of the starch that must be added each day to keep the protozoa alive is important. Some commercially available starches, e.g. potato, contain a large proportion of grains that are too large to be engulfed by many protozoal species and give poor growth, not only because the protozoa

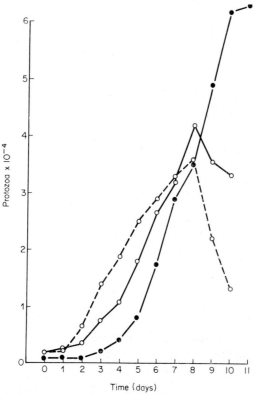

FIG. 2. Growth of *Entodinium caudatum* and *E. simplex* from a small inoculum. *E. caudatum* was grown from an inoculum of 900 protozoa/ml in caudatum-type salts containing 10% prepared fresh rumen fluid with the daily addition of 0·25 mg rice starch (closed circles). *E. simplex* was grown from an inoculum of 1800 protozoa/ml in simplex-type salts containing 10% autoclaved rumen fluid with the daily addition of 0·5 mg wholemeal flour/ml (open circles and solid line) or 1·0 mg wholemeal flour/ml (open circles and broken line).

lack a source of food but because the large grains are fermented in the medium. Where possible a starch with small grains e.g. rice starch, is used and *Entodinium caudatum* grows well on this material. Other species such as *Entodinium simplex* and *Epidinium ecaudatum caudatum* will only grow on a crude milled grain such as wholemeal flour. This cannot be replaced by rice starch or wheat starch unless they are supplemented with washed bran (Coleman 1969a).

As mentioned above the quantity of starch added each day is also important and for an established culture growing in the presence of auto-claved rumen fluid 0·12 mg rice starch/ml or 0·25–0·5 mg wholemeal flour/ml added daily gives satisfactory growth. Double these quantities often stimulates the growth rate and produces higher population densities but in cultures that are not diluted regularly it also produces a rapid decline in numbers of protozoa once the maximum population density is passed. Growth curves for *Entodinium simplex* showing some of these effects are shown in Fig. 2. As this is associated with heavy bacterial growth, death of cultures sometimes occurs under these conditions. For the long-term maintenance of protozoal cultures it is safer to dilute the cultures regularly with an equal amount of fresh medium and to add the smaller quantities of starch each day. However, the optimum amount of starch to add each day does depend on the type of rumen fluid used and in the presence of fresh rumen fluid the optimum is about double that in the presence of autoclaved rumen fluid.

Dried grass

This is essential for the prolonged cultivation of *Entodinium caudatum* and *E. simplex* at least. The former organism is known to engulf chloroplasts as has been shown by examination of thin-sections in the electron micro-scope (Coleman and Hall 1969). However cultures of these protozoa will grow normally for more than a week in the absence of dried grass although a few mg is added routinely to all cultures each day. The grass only supported protozoal growth if it had been dried below 60° and sterilization in the dry state at 140° resulted in death of the protozoa.

Temperature

The optimum temperature for all species is 37° to 40° and they will not grow above 42° or below 35°. Growth of *Entodinium caudatum* at 41° resulted in distorted protozoa that did not separate normally during binary fission and grew in chains of up to five organisms. The protozoa will survive for short periods at lower temperatures and if a culture of *E. caudatum* is

kept at room temperature for 12 h and then returned to an incubator at 39°
a few protozoa will be found alive 12 h later.

Initial Isolation and Daily Maintenance of Protozoal Cultures

General considerations

Starch

The most important single factor in the growth of protozoal cultures from
single organisms is the amount of starch that is added each day. Any
excess starch that is not engulfed by the protozoa is fermented by bacteria
in the bottom of the tube. As the protozoa are also at the bottom of the
tube it is probable that local unfavourable conditions are produced which
kill the protozoa. Conversely, when the protozoa begin to multiply it
is essential to increase the amount of starch that is added each day to
prevent them dying of starvation. The correct amount can only be determined by trial and error but approximately 0·5 mg wholemeal flour
should be the optimum amount for an inoculum of 1–5 protozoa in 10 ml
medium.

Once the culture is established and growing well the daily addition of
starch is still the most important factor in its maintenance, as none of the
protozoal species cultivated by the author will survive unless fresh starch is
added regularly every day. With some species e.g. *Entodinium caudatum*
and *Epidinium ecaudatum caudatum* a single gap of 48 h is not fatal but none
can be maintained by feeding only every 48 h.

Medium Replacement

Some protozoal species will grow well from a single organism for a time
with the daily addition of wholemeal flour and then individual protozoa
become much less motile and begin to die. Such cultures can often die
completely in 24 h unless the supernatant fluid is replaced by fresh medium.
It may be necessary to do this several times before the population density
is such that the culture can be diluted with an equal volume of fresh
medium. For all the species the population density obtainable increases
steadily over the first few months in culture and the values quoted in Table
1 are the maxima obtainable in established cultures.

Contamination

As it is usual when isolating individual protozoa to inoculate a series
of tubes care must be taken when feeding the cultures each day not to
transfer protozoa from one tube to another. If this precaution is not taken it
is easy for a vigorously growing contaminant protozoon such as *Entodinium*

caudatum to be transferred from one tube, into which it was accidentally inoculated originally, to all the others in a batch. The necessary precautions include, always using the same rubber bung for a tube, flaming the pipette used to gas the tubes between each one and never allowing the pipette used to add the starch suspension to touch the side of a tube.

The details of the methods used to isolate and maintain individual species are given below.

Entodinium caudatum

Initial isolation

Entodinium caudatum can be isolated in pure culture (in the presence of bacteria) by enrichment in caudatum-type medium in the presence of 10% prepared fresh rumen fluid with the daily addition of 0·5 mg rice starch/ml and dried grass.

A 10% inoculum of crude rumen contents is used and initially the culture is centrifuged and the supernatant fluid replaced by fresh medium each day. After 2–3 weeks the other protozoal species disappear and the population density of *E. caudatum* increases. The culture is then diluted with an equal volume of fresh medium and this process is repeated once a week for approximately 6 months after which time the protozoa will be growing fast enough for the culture to be diluted twice a week. Once the *E. caudatum* has begun to grow, the daily replacement of the culture supernatant can be omitted except on days when the culture is diluted; daily feeding with rice starch must, of course, continue. It is also possible to omit the centrifugation before removal of the supernatant on "dilution-days" as most of the protozoa are already present on the bottom of the tube. After some months it is possible to replace the prepared fresh rumen fluid by autoclaved rumen fluid without decreasing the growth rate, but this change should be associated with a decrease in the amount of rice starch added each day to 0·1–0·2 mg/ml.

In the initial isolation of *Entodinium caudatum* (Coleman 1958, 1960) the addition of 50 µg chloramphenicol/ml was stimulatory to protozoal growth. Subsequent attempts to leave chloramphenicol out of the medium have always resulted in the death of the culture in a few weeks and it has therefore always been added to the medium. However, in more recent attempts to grow other protozoal species it has proved very easy to grow *Entodinium caudatum* organisms that have been introduced accidentally into medium lacking chloramphenicol. In the author's experience although it has often proved difficult to grow a certain protozoal species for the first time it has subsequently been comparatively easy to grow a culture from an inoculum of only a few protozoa. The reason for this is not clear. It may be an unconscious improvement in handling technique or, as no attempt is made to sterilize the food materials, it may be that it takes time for a suitable bacterial flora to become established. White (1969) showed that *Entodinium caudatum* cultures maintained *in vitro* for 6 years contained as their principal bacteria, two Gram-negative bacilli, *Klebsiella aerogenes* and *Proteus mirabilis* which are not normally present in appreciable numbers in the rumen. The exact relationship between protozoa and their associated bacteria is not clear but it is possible that only certain bacterial species will grow in the medium and be compatible with protozoal growth. If these bacteria are not normally present in the air or food materials then it may take some time for them to be inoculated by chance into the protozoal cultures. However once these bacteria are present in protozoal cultures they will become distributed throughout the laboratory and their subsequent inoculation into fresh medium will occur readily. As a result subsequent attempts to grow the protozoa directly from the rumen will be greatly facilitated.

Maintenance of established culture

Entodinium caudatum grows most rapidly at population densities of 10,000 to 30,000/ml and the cultures are usually maintained by dilution with an equal volume of fresh caudatum-type medium twice a week. In practice the following procedure is adopted. The centrifuge tubes in which the protozoa are grown are removed from the incubator one at a time and the top two-thirds of the liquid removed. As the protozoa are found principally as a loose pellet at the bottom of the tubes this procedure does not remove many of them, especially if the liquid is removed before convection currents set up by cooling disturb the pellet. The liquid that has been removed is replaced by fresh caudatum-type medium, the tube gassed, mixed and half poured into a new tube. Each is then made up to the original volume with fresh medium, a few mg of dried grass added and the tubes gassed with $95\% \ N_2 + 5\% \ CO_2$ before sealing with a rubber bung. On days when the cultures are not diluted, rice starch (to give a final concentration of 0·12 mg/ml) and a few mg of dried grass are added to each tube before gassing with $95\% \ N_2 + 5\% \ CO_2$ and sealing the tube. It is possible to maintain *E. caudatum* for about 3 weeks without changing the supernatant fluid but after one week the population density declines steadily.

Entodinium simplex

Initial isolation

Entodinium simplex can be isolated in simplex-type medium containing 10% (v/v) prepared fresh rumen fluid with the daily addition of wholemeal flour. As *E. simplex* is the smallest rumen ciliate protozoon, it is separated from the other protozoa by differential centrifugation. The exact conditions vary with the rumen contents used, but 2–3 min at 500–1000 **g** will usually sediment all the largest protozoa and leave 10–20 *E. simplex*/ml in the supernatant fluid. As an inoculum of only a few protozoa is required to initiate a culture only a drop of the supernatant fluid has to be added. The safest method of ensuring that only *E. simplex* is added is to examine under a microscope the actual material that is to be used for inoculation. The cultures are first established in 15 ml centrifuge tubes containing 10 ml medium to which were added each day 0·1 ml 1·5% (w/v) suspension of wholemeal flour plus about 1 mg dried grass. After 10 to 14 days there are usually 50–100 protozoa/ml and a week later the culture is diluted with an equal volume of fresh medium and transferred to a 50 ml centrifuge tube. The quantity of 1·5% wholemeal flour suspension added daily is gradually increased to 0·5 ml and then doubled to 1·0 ml. The cultures are diluted with an equal volume of fresh medium once a week for a month, after which time this is increased to twice each week. About two months after the initial isolation it is possible to replace the prepared fresh rumen fluid in the medium by autoclaved rumen fluid without decreasing the protozoal growth rate. It is important to keep a constant check on these cultures for the presence of any caudate protozoa as these can easily lose their tails and come to resemble large *E. simplex*. A culture of *E. simplex* can only be considered to be pure if none of the protozoa have ever had caudal spines.

Maintenance of established culture

Twice a week, the contents of each tube are mixed and half poured into a clean tube. Each is made up to the original volume with simplex-type medium and 0.3 mg wholemeal flour/ml and a few mg dried grass added. The tubes are gassed with CO_2 and sealed with a rubber bung. On days when the culture is not diluted 0.3 mg wholemeal flour/ml and a few mg dried grass are added before gassing and sealing the tubes.

Epidinium ecaudatum caudatum

Initial isolation

Epidinium ecaudatum caudatum grows well in the same medium as that used for *Entodinium simplex* but can only be separated from fresh rumen contents or a mixed culture by picking out single organisms with a micro-manipulator. The epidinia resist exposure to oxygen better than the other Entodiniomorphid protozoa that have been cultivated *in vitro* and have proved very easy to isolate in pure protozoal culture (in presence of bacteria) by this means. The initial isolation is made in 10 ml medium and then the volume of the culture and the quantity of wholemeal flour/ml added daily are increased as described for *E. simplex*. Unlike the entodinia *Epidinium ecaudatum caudatum* grows rapidly from the beginning and once the population density exceeds 300/ml the cultures can be diluted with an equal volume of fresh medium twice each week.

Maintenance of established cultures

Twice a week, the contents of each tube are mixed and half poured into a clean tube. Each is made up to the original volume with dilute simplex-type medium and 0·3 mg wholemeal flour/ml and a few mg dried grass added. The tubes are gassed with CO_2 and sealed with a rubber bung. On days when the culture is not diluted 0·3 mg wholemeal flour/ml and a few mg dried grass are added before gassing and sealing the tubes.

Polyplastron multivesiculatum

Initial isolation

As stated above, the proportion of *Polyplastron multivesiculatum* in a culture or in crude rumen contents can be increased by growth in Hungate-type salts (from which free oxygen has been removed), in the absence of rumen fluid or cysteine, to which wholemeal flour and dried grass are added daily. However, it has proved difficult to obtain growth of single protozoa taken from this medium and inoculated into the same medium or into simplex or caudatum-type medium containing 10% prepared fresh rumen fluid. In contrast single organisms (from an enrichment culture or from the rumen) inoculated into a culture of *Epidinium ecaudatum caudatum* (growing in simplex-type medium) always grew well and divided at least every 24 h until all the epidinia had disappeared. At least some of the epidinia are engulfed by the *Polyplastron multivesiculatum* but it is uncertain if this is the only reason for their disappearance. Provided that fresh epidinia (added as a culture of *Epidinium ecaudatum caudatum*) are added every 2–3 days the cultures can be maintained by dilution with fresh simplex-type medium containing 10% fresh rumen fluid at least once a week. After the culture has been established for about two months *Polyplastron multivesiculatum* will grow if inoculated into caudatum-type medium containing 10% fresh rumen fluid but still only if epidinia are added regularly. However, after a few weeks' growth under these conditions the epidinia can be omitted and *P. multivesiculatum* will grow in their absence. Any residual epidinia that are present are soon engulfed by the *P. multivesiculatum* organisms. Unfortunately such cultures often die inexplicably and, as might be expected from the above observations, the maintenance of *P. multivesiculatum* in the absence of epidinia is best carried out in Hungate-type salts with dilution of the culture with fresh medium once a week. At the time of writing the optimum conditions have not been elucidated but growth is obtained with the daily addition of 0·1 mg wholemeal flour/ml and a few mg of dried grass. Irrespective of the exact growth conditions used, better and more reproducible growth was obtained in conical flasks half filled with medium rather than in centrifuge tubes.

Maintenance of established cultures

Up to the time of writing established cultures of *Polyplastron multivesiculatum* have been maintained in conical flasks under two conditions.

(a) Under 100% CO_2. Twice a week the contents of the flasks are mixed, and half poured into a clean flask. Each is made up to the original volume with simplex-type medium containing 10% prepared fresh rumen fluid and 0·6 mg wholemeal flour/ml, and a few mg dried grass and 10 epidinia (added as a culture of *Epidinium ecaudatum caudatum*)/ml are added. The flasks are gassed with CO_2 and sealed with a rubber bung. On days when the culture is not diluted 0·3 mg wholemeal flour/ml, dried grass and epidinia are added as before.

(b) Under 95% N_2 + 5% CO_2. As for (a) except that caudatum-type medium containing 10% fresh rumen fluid is used and the flasks are gassed with 95% N_2 + 5% CO_2.

Although the *Polyplastron multivesiculatum* in the author's cultures were isolated on simplex-type medium the steady state population densities are 15/ml under condition (a) and 50/ml under condition (b).

A summary of these methods and some information on an *Ophryoscolex* sp., probably *O. purkynei*, is shown in Table 2.

The author has also maintained *Entodinium bursa* alive *in vitro* in the presence of *Entodinium caudatum* for over a year on simplex-type medium + 0·02% glucose with the daily addition of wholemeal flour and dried grass. The ciliates have grown from an initial inoculum of crude rumen contents. The culture supernatant fluid has been replaced three times each week and the culture diluted with fresh medium once a fortnight. Unfortunately attempts to isolate *E. bursa* in the absence of other protozoa have

TABLE 2. Summary of the methods used for the isolation and maintenance of Entodiniomorphid protozoa

Protozoon	Years cultivated in vitro	Initial isolation	Starch source	Optimum-proportion of CO_2	Necessity for other protozoa	Method of increasing culture*
Entodinium caudatum	10	Enrichment	Rice starch	5%	None	Replace medium and dilution
Entodinium simplex	3½	Differential centrifugation	Wholemeal flour	100%	None	Dilution
Epidinium ecaudatum caudatum	¾	Single cell	Wholemeal flour	100%	None	Dilution
Polyplastron multivesiculatum	½	Single cell	Wholemeal flour	5%	None	Dilution
				100%	epidinia	Dilution
Ophryoscolex sp.	⅓	Single cell	Wholemeal flour	5%	None	Dilution

* This refers to the twice-weekly method of increasing the volume of the culture. Dilution = add an equal volume of fresh medium.

been unsuccessful and in view of the known predatory habits of this protozoon it may have an obligate requirement for other entodinia as a source of food.

Attempts to increase growth rate

It is possible to grow the entodinia, at least, with a much shorter generation time by dilution of the culture every day or every other day rather than twice a week. Unfortunately as the generation time is decreased the number of protozoa also decreases and steady state population densities of *Entodinium simplex* of 1100, 21,400 and 26,900/ml were obtained when cultures were diluted every day, every other day and twice a week respectively.

Growth from a small inoculum

Although most of the Entodiomorphid protozoa described above have been maintained routinely at comparatively high population densities so that with e.g. *Entodinium caudatum* the numbers never dropped below 10,000/

ml, they will also grow from single organisms (see "Initial Isolation") or from an initial density of 100–2000/ml. Unfortunately there is usually an appreciable lag of up to three days before growth begins from these small inocula (Fig. 1) and occasionally all the protozoa die.

Acknowledgements

The author wishes to record his thanks to Miss C. M. Tebbutt, Miss J. M. How, Miss J. Pearson, Miss J. King, Miss E. V. Holgate, Miss M. Sale, Mrs B. C. Barker and Miss J. I. Davies who between them have cherished the protozoa over the past fifteen years and without whose assistance this review could never have been written.

References

CLARKE, R. T. J. (1963). The cultivation of some rumen oligotrich protozoa. *J. gen. Microbiol.*, **33**, 401.

COLEMAN, G. S. (1958). Maintenance of oligotrich protozoa from the sheep rumen *in vitro*. *Nature, Lond.*, **182**, 1104.

COLEMAN, G. S. (1960). The cultivation of sheep rumen oligotrich protozoa *in vitro*. *J. gen. Microbiol.*, **22**, 555.

COLEMAN, G. S. (1969a). The cultivation of the rumen ciliate *Entodinium simplex*. *J. gen. Microbiol.*, **57**, 81.

COLEMAN, G. S. (1969b). The metabolism of starch, maltose, glucose, and some other sugars by the rumen ciliate *Entodinium caudatum*. *J. gen. Microbiol.*, **57**, 303.

COLEMAN, G. S. & HALL, F. J. (1969). The engulfment of bacteria and other particulate matter by the rumen ciliate *Entodinium caudatum*. *Tissue and Cell*, **1**, 607.

GUTIERREZ, J. & DAVIS, R. E. (1962). Culture and metabolism of the rumen ciliate *Epidinium ecaudatum* (Crawley). *Appl. Microbiol.*, **10**, 305.

HUNGATE, R. E. (1942). The culture of *Eudiplodinium neglectum*, with experiments on the digestion of cellulose. *Biol. Bull. mar. biol. Lab., Woods Hole*, **83**, 303.

HUNGATE, R. E. (1943). Further experiments on cellulose digestion by the protozoa in the rumen of cattle. *Biol. Bull. mar. biol. Lab., Woods Hole*, **84**, 157.

HUNGATE, R. E. (1955). Mutualistic intestinal protozoa. In *"Biochemistry and Physiology of Protozoa"*. (S. H. Hutner and A. Lwoff, eds) Vol. II p. 159. New York and London: Academic Press.

JARVIS, B. D. W. & HUNGATE, R. E. (1968). Factors influencing agnotiobiotic cultures of the rumen ciliate *Entodinium simplex*, *Appl. Microbiol.*, **16**, 1044.

MAH, R. A. (1964). Factors influencing the *in vitro* culture of the rumen ciliate *Ophryoscolex purkynei* Stein. *J. Protozool.*, **11**, 546.

TOMPKIN, R. B., PURSER, D. B. & WEISER, H. H. (1966). Influence of rumen fluid source upon establishment and cultivation *in vitro* of the rumen protozoon *Entodinium*. *J. Protozool.*, **13**, 55.

WHITE, R. W. (1969). Viable bacteria inside the rumen ciliate *Entodinium caudatum*. *J. gen. Microbiol.*, **56**, 403.

Anaerobic Organisms from the Human Mouth

G. H. Bowden and J. M. Hardie

Dental Bacteriology and Biochemistry Department,
The London Hospital Medical College, London, England

This chapter describes some of the methods which can be used to isolate and identify anaerobic bacteria from the mouth. The isolation procedures and most of the identification methods described have been used successfully in our laboratory. If undue emphasis appears to have been laid on particular genera, this reflects the personal interests of the authors. There is more information available in the literature about some groups of organisms than others and areas of oral microbiology still exist that are relatively unexplored. Wherever possible, references relevant to the organisms under consideration have been included.

Sampling

The mouth supports an extremely varied flora which includes obligate aerobes and facultative organisms. While representative species of these groups can be isolated from most areas of the mouth, certain sites favour particular genera.

Oral samples can be divided into the following types:

(1) Whole saliva (pure saliva collected by cannulation of individual salivary glands is usually sterile).

(a) Stimulated. Produced by chewing inert material such as wax or a rubber band.

(b) Unstimulated. Collected directly from the mouth.

(2) Microbial aggregations on tooth surfaces (Dental Plaque).

(a) Gingival plaque (i.e. adjacent to the gum margin).

(b) Interstitial plaque (i.e. from areas between adjacent teeth).

(c) Plaque from other surfaces (i.e. buccal, lingual, occlusal).

(3) Soft tissue areas.

(a) Gingiva.

(b) Tongue.

(c) Cheek and other areas.

Saliva

Saliva supports a varied flora (Table 1), the composition of which is

TABLE 1. Viable* organisms in plaque and saliva (Expressed as a percentage of total viable* organisms)

	Duration of growth (days) in Plaque					Saliva
	1 Day	3 Days	5 Days	7 Days	14 Days	
Streptococci	94–99·0	56–99·0	39–99·0	47–95·0	74–83·0	32–79·0
Lactobaccilli	0– 1·0	0–43·0	0–54·0	0–47·0	0–15·0	0–17·0
Neisseria	0·5–5·0	0–10·0	0– 2·0	0– 0·5	0– 0·9	6·8–65·0
Nocardia (Rothia)	0–0·19	0–0·5	0–18·4	0–0·28	0– 1·3	0–11·2
Veillonella	0– 1·1	0–14·5	0–37·0	0– 7·0	0– 4·3	0– 9·3
Fusiform	0	0–0·05	0–0·56	0– 2·8	0– 0·1	0–0·48
B. melaninogenicus	0	0– 4·9	0–12·0	0– 3·0	0	0–24·7
L. buccalis	0	0–0·004	0–0·003	0–0·003	0– 3·4	0–0·019
Anaerobic filaments including *Actinomyces* and *L. aerogenes.*	0	0–0·05	0–0·62	0·05–11·4	9·6–13·1	0–1·37

Results from Slack and Bowden (1965).

 * The table indicates the numbers of viable organisms which are present in Dental Plaque as it develops on the tooth surface. The numbers of days refer to the period of time that the plaque was allowed to develop prior to a sample being taken. The saliva counts are included for comparison and represent an average of the numbers of viable organisms which can be found in a saliva sample at any given time. As plaque samples cannot be taken with any accuracy the figures are expressed as a percentage of the total viable organisms in the sample.

relatively stable when compared to the flora of dental plaque. Several anaerobes, especially *Veillonella, Fusobacteria* and *Bacteroides* spp, can be isolated from most samples. In most cases inoculation of salivary centrifuged deposit is more successful than plating saliva direct. Saliva may be collected by dribbling into a sterile bottle or removal from the mouth with a Pasteur pipette. If the subject forces some saliva between the teeth a few times before collection a higher concentration of organisms in the specimen can be attained.

Dental Plaque

(For illustration see Fig. 1.) This consists of a mass of microorganisms and their products which develops on a tooth surface. Clearly visible masses occur within 24 h of thorough dental prophylaxis, especially in the gingival

FIG. 1. Dental plaque (Gram's stain; × 800). Large Gram-negative filamentous organisms grow perpendicular to the tooth surface (bottom of picture); Gram-positive organisms can be seen distributed singly and in masses between these filaments.

o

areas, and these can be demonstrated clinically with disclosing solutions. The composition of plaque varies over the tooth surface so that areas within a few millimetres of each other will have a different flora. The majority of samples, however, are taken from a relatively large area and several general observations can be made. Streptococci are almost invariably the dominant viable organisms, whilst *Veillonella, Bacteroides, Fusobacteria* and *Actinomyces* spp are present in most samples (Ritz, 1967). Filaments resembling *Leptotrichia* spp are clearly visible on microscopic study, but viable counts of these are usually low. Plaque from protected areas, such as interstitial plaque, is mainly filamentous whilst gingival plaque tends to contain high numbers of fusobacteria, bacteroides and spirochaetes. Dental calculus, or tartar, is dental plaque that has become calcified.

Samples of plaque should be taken with a sterile metal instrument which can be used to scrape the tooth surface. Cotton swabs and soft wire loops are not suitable.

Soft tissue areas

Cotton swabs can be used to sample these areas. The area commonly sampled for anaerobes is the crevice or pocket that exists between the gingiva and the tooth surface. Sterile filter paper triangles or fine capillary tubes can be used for this purpose and these are gently inserted into the crevice. The dorsum of the tongue carries a large and characteristic flora and may well be the principal source of many of the organisms that are found in saliva.

For separation and enumeration of different species from a sample a serial dilution technique can be used with the appropriate media (Slack and Bowden, 1965). Dilutions of 1:10, 1:100, 1:500, 1:1000 and 1:5000 can be used with a volume of 0·02 ml spread on to each plate. The dilutions should be prepared in a reducing medium to protect anaerobic organisms. The following diluent has been found suitable:

Tryptone (Difco)	1·0 g
Yeast Extract (Difco)	0·5 g
Glucose	0·1 g
Cysteine hydrochloride	0·1 g
Bovine Serum (Oxoid)	2·0 ml
Distilled water	100 ml

The medium is sterilized by filtration

Plates should be prepared fresh, and if they have to be stored they should be kept in a reduced atmosphere in anaerobic jars. Incubation should be carried out under 95 % H_2 + 5 % CO_2 for 7 days.

Growth should be examined under a plate microscope. Low dilutions on non-selective media will have large numbers of colonies, but with care colonies can be selected and subcultured using a fine straight wire. Selective media are useful for isolation, but viable counts of specific organisms will always be lower on these media than on non-selective media.

The more elaborate anaerobic isolation methods which have been applied to intestinal flora have not been used for oral samples. However, using the methods outlined above it is still possible to isolate organisms which cannot be identified with any certainty at present.

Actinomyces

The mouth is the natural habitat of several *Actinomyces* spp, including the human pathogen *Actinomyces israeli*. Cultural studies on the oral flora by Howell, Murphy, Paul and Stephan (1959) and Howell, Stephan and Paul (1962) and fluorescent antibody examinations (Snyder, Bullock and Parker, 1967; Baboolal, 1968; Gerencser, Landfried and Slack, 1969) have shown that actinomyces are present in most oral samples. A numerical taxonomic study of the oral Gram-positive filaments including *Actinomyces* spp has been made (Melville, 1965).

Classification of *Actinomyces* has been confused in the past, but serological and cell wall examinations have allowed the separation of several species. Those currently recognized are *A. israeli*, serotypes 1 and 2 (Brock and Georg, 1969), *A. ericksonii* (Georg, Robertstadt, Brinkman and Hicklin, 1965), *A. naeslundi* (Thompson and Lovestedt, 1951), *A. odontolyticus* (Batty, 1958). *A. bovis* (Harz, 1878; Erickson, 1940) and *A. viscosus* (previously called *Odontomyces viscosus*; Howell, 1963; Howell, Jordan, Georg and Pine, 1965; Gerencser and Slack, 1969; Georg, Pine and Gerencser, 1969). The organism previously named *Actinomyces propionicus* (Buchanan and Pine, 1962; Gerencser and Slack, 1967) has recently been proposed as a new genus, *Arachnia propionica* (Pine and Georg, 1969).

Of these species, *A. israeli* (both serotypes), *A. naeslundi*, *A. odontolyticus*, *A. viscosus* and *Arachnia propionica* have been isolated from the mouth. *A. israeli* is microaerophilic or anaerobic, but can best be isolated under strictly anaerobic conditions. The other oral *Actinomyces* spp are more oxygen tolerant.

Isolation

Selective media have been used for the isolation of actinomyces from oral (Howell *et al.*, 1959) and clinical samples (Blank and Georg, 1968). None of these media have been very effective for oral samples in our hands, but this may be due to variations in the sources of chemicals and dyes used. We have found the most successful technique to be 1:100, 1:1000 and 1:5000 dilutions of interstitial plaque plated on blood agar and incubated anaerobically at 37° for 7 days.

Typical colonies are picked off and those showing Gram-positive, pleomorphic, branching organisms are selected for further tests. Colonies of *A. israeli* are usually 1–2 mm in diam and of the rough, heaped, "bread crumb" type. *A. naeslundi* colonies are similar but more smooth, the colonies of *A. odontolyticus* are 2 mm, smooth and reddish-brown after 4–7 days' incubation. Mature *A. viscosus* colonies are soft or mucoid, cream-white, 2 mm in diam, often with radial or concentric striations.

Preliminary separation of species can be made upon morphology and atmospheric requirements. Further differential tests are shown in Table 4. Serological identification is possible using gel diffusion of culture super-natant antigens (King and Meyer, 1963; Georg, Roberstad and Brinkman, 1964) or fluorescent antibody technique (Lambert, Brown and Georg, 1967; Slack, Landfried and Gerencser, 1969). Cell wall studies provided one of the earliest clear differentiations of species (Cummins and Harris, 1958; Cummins, 1962) and are very useful in identification. Such studies are especially valuable in the differentiation of actinomyces from "anaerobic diphtheroids" which often contain diaminopimelic acid and arabinose (Boone and Pine, 1968).

Other oral Gram-positive filaments might be confused with the actino-myces. These include *Rothia, Bacterionema* and some corynebacteria, which are all however, catalase positive and have distinctive cellular morphology. *Rothia* spp branch profusely and fragment into coccal forms (Davis and Freer, 1960; Georg and Brown, 1967; Roth and Thurn, 1962).

Bacterionema matruchotii (previously called *L. dentium*) has the charac-teristic "whip handle" cell and commonly branches (Baird-Parker and Davis, 1958; Gilmour, Howell and Bibby, 1961). Oral corynebacteria very seldom branch and their short cells are often septate, commonly with club-ended forms. Cell wall stains (Bisset and Davis, 1960) are useful in morphological studies of these organisms (Fig. 2). The character-ization of *Arachnia propionica* is dealt with by Gerencser and Slack (1967).

All of the oral actinomyces grow well on blood agar and brain heart infusion agar. The best fluid medium is that of Ajello, Georg, Kaplan and Kaufman (1963) which can be obtained commercially from Baltimore Biological Laboratories. Most species grow well in this medium in static culture, although growth of *A. odontolyticus* may be increased by aeration and shaking.

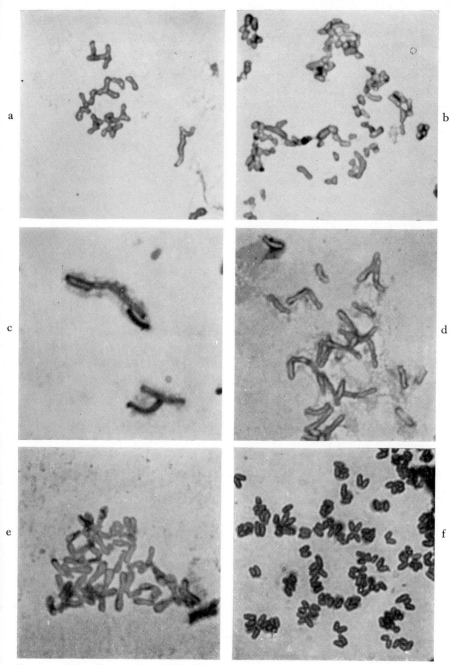

FIG. 2. Cell wall stain (× 1600). (a), *Rothia* spp, 24 h culture prior to fragmentation; (b), *Rothia* spp, 72 h culture showing fragmentation into shorter rod and coccal forms; (c), *B. matruchotii*, "whip handle" cell; (d), *A. naeslundi*, 48 h culture from blood agar; (e), oral coryneform organism, showing club forms, and (f), oral coryneform organism, showing typical arrangement and an occasional septate cell.

TABLE 4. Reactions and cell wall components of *Actinomyces* species

Reactions	A. israeli/1	A. israeli/2	A. naeslundi	A. Odontolyticus	A. ericksonii	A. bovis	A. viscous
Glucose	acid	acid	acid	acid	acid	acid	acid
Salicin	variable	variable	variable	—	acid	acid	acid
Raffinose	acid	—	usually +	—	acid	—	acid
Arabinose	acid	variable	—	variable	acid	acid	—
Mannite	variable	usually +	—	variable	acid	—	variable
Cellobiose	usually +	acid	—	variable	acid	—	—
Xylose	acid	acid	variable	—	acid	variable	acid
Inositol	acid	usually +	usually +	—	acid	acid	—
NO₃ Reduction	usually +	usually +	usually +	usually +	—	—	—
NO₂ Reduction	—	weak +	—	—	+	+	+
Starch Hydrolysis	weak +	—	—	—	—	—	+
Catalase	—	—	—	—	+	—	usually +
Growth O₂	usually —	usually —	usually +	+	—	slight	+ +
Growth 10 % CO₂	slight	slight	+ +	+ +	—	+	—
AnO₂ + CO₂	+	+	+ + +	+ + +	+ +	+ + +	+ +
Aesculin	+	+	+	+	+	+	+
Cell Wall Components							
Glucose	—	—	+	+	—	+	+
Galactose	+ or —	+	—	—	+	—	trace
Mannose	—	+ or —	trace	—	+ or —	—	trace
Rhamnose	+	+	+	+	—	+	+
Deoxy Hexose	—	—	—	—	—	—	—
Glucosamine	+	+	+	+	+	+	+
Diaminopimelic acid	—	—	—	—	—	—	—
Ornithine	+	+	+	+	+	—	+

Compiled from published work by: Batty (1958), Boone and Pine (1968), Brock and Georg (1969), Cummins and Harris (1958), Cummins (1962), DeWeese, Gerencser and Slack (1968), Georg, Roberstadt and Brinkman (1964), Georg *et al.* (1965), Gerencser and Slack (1969), Pine and Boone (1967) and Slack, Landfried and Gerencser (1968).

Reliable fermentation reactions may be obtained using either Actinomyces Fermentation Broth (B.B.L.) or the Medium given below:

Proteose Peptone (Oxoid)	2·0 g
Yeast Extract (Difco)	0·5 g
Na$_2$HPO$_4$ (anhydrous)	0·02 g
Cysteine HCL	0·01 g
Phenol Red	0·002 g
Distilled water	100 ml

Autoclave at 121° for 15 min.

Add carbohydrate to a concentration of 1% as a filter sterilized solution, and 0·5% sterile bovine serum (Oxoid).

Leptotrichia

The genus *Leptotrichia* was ill-defined until the late nineteen fifties and was not included in the 7th Edition of Bergey's Manual (1957). One oral species, *Leptotrichia buccalis*, is generally accepted whilst Theilade and Gilmour (1961) have described an oral filament which may be related to it. Several recent studies have been made on this group, including those of Hamilton and Zahler (1957), Davis and Baird-Parker (1959 *a*, *b*), Baird-Parker and Davis (1958) and Kasai (1961, 1965).

L. buccalis is a fusiform filamentous organism forming individual cells up to 15 μ in length and 0·8–1·5 μ wide. The cells commonly occur in pairs, with the adjacent ends flattened, but can form chains up to 100 μ long (Fig. 3). Early cultures, up to 16 h, show Gram-positive cells and Gram-negative cells with positive-staining granules. Older cultures are almost

TABLE 2. Reactions* of *Leptotrichia buccalis* and "*Leptotrichia aerogenes*"

	L. buccalis	L. aerogenes
Catalase	—	—
Glucose	acid	acid
Lactose	acid	acid
Sucrose	acid	acid
Mannite	—	—
Raffinose	—	acid
Galactose	variable	acid
Arabinose	—	—
Indole	—	+
Gas Production	—	+

* Data taken from Hofstad (1967*a*) and Baboolal (1969).

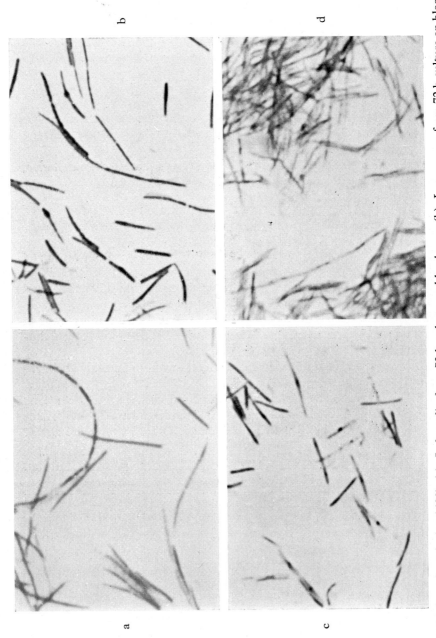

Fig. 3. Gram's stain (× 1600). (a), *L. buccalis*, from 72 h culture on blood agar; (b), *L. aerogenes*, from 72 h culture on blood agar; (c), *Fusobacterium*, short cells with granules, and (d), *Fusobacterium*, longer cell forms.

invariably Gram-negative. Branching cells are very seldom seen and no motile strains have been described.

Colonies are distinctive (Fig. 4) especially when grown on the basal media described by Kasai (1961) and Baird-Parker (1957). After 72 h incubation they are smooth, colourless, 2–3 mm in diam, rhizoid with striations and

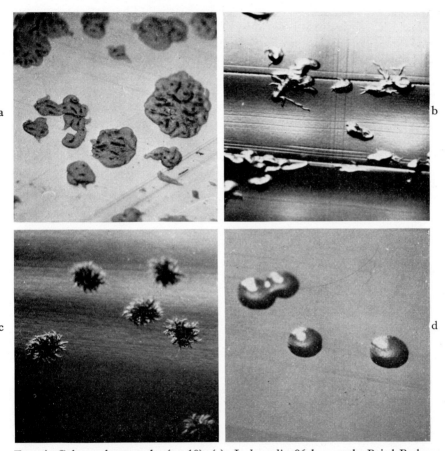

FIG. 4. Colony photographs (× 10). (a), *L. buccalis*, 96 h growth, Baird-Parker medium base, convoluted type of colony; (b), *L. buccalis*, 48 h growth blood agar, showing the early rhizoid type of colony; (c), "*L. aerogenes*" colony, 48 h growth on blood agar, and (d), smooth type of *Fusobacterium* colony 48 h growth on blood agar.

are described as "medusa head". The organisms are anaerobic on initial isolation but adaptation to aerobic growth has been reported. All strains are active against several sugars but none produce indole or gas (Table 2).

Cell wall studies have been made and suggest a Gram-positive structure although endotoxins can be extracted from the organisms (Gustafson, Kroeger and Vaichulis, 1966). Electron microscope studies show similarities between *L. buccalis* and typical Gram-negative cell structure (Hofstad and Selvig, 1969).

The second proposed species described by Theilade and Gilmour (1961) resembles *L. buccalis* in its Gram-staining reaction. The filaments are up to 20 μ in length and 0·5–1·0 μ wide, often with swellings along their length, and generally less fusiform in shape than *L. buccalis* (Fig. 3). Colonies are 2–4 mm in diam after 7 days, non-pigmented with a veined surface and finely rhizoid edge (Fig. 4). The organisms are strictly anaerobic and differ from *L. buccalis* in the production of indole and gas, other reactions being similar (Table 2).

The name *L. aerogenes* has been proposed for this species (Hofstad, 1967a), although it is sometimes referred to as *L. ginsii*. *L. aerogenes* can be differentiated (Table 3) from *L. buccalis* by its cell wall components

TABLE 3. Wall components of *Leptotrichia buccalis* and "*Leptotrichia aerogenes*"

	L. buccalis	L. aerogenes
Alanine	+	+
Glutamic acid	+	+
Diaminopimelic acid	+	+
Lysine	+	+
Glycine	+	+
Serine	+	+
Valine	+	+
Aspartic acid	+	+
Threonine	+	+
Glucose	+	+
Rhamnose	—	+
Heptose	—	+

Includes results from Hofstad (1967a, b) and Baboolal (1969).

(Hofstad, 1967b; Baboolal, 1969), and electron microscope studies indicate a Gram-negative wall structure (Hofstad, 1969b).

Organisms closely resembling leptotrichia can be seen in most microscopic examinations of dental plaque. Sections of plaque show that the filaments are arranged perpendicular to the tooth surface (Fig. 1), both *Leptotrichia* spp can be isolated from samples of dental plaque especially those from the gingival area. Plates should be inoculated with undiluted plaque, as the viable counts of these organisms usually represent only 5–7% of the total plaque flora.

Two selective media which are useful for isolation of *L. buccalis* are those of Omata and Disraely (1956) and Baird-Parker (1957). Baird-Parker medium is selective for *L. buccalis* whilst the medium of Omata and Disraely also supports the growth of fusobacteria. In both media the selective agents are dyes and antibiotics. Individual batches of dyes must be checked for their suitability and this is especially true of ethyl violet in Baird-Parker medium. The basal media are useful for isolation once the colonies can be recognized. Although they are non-selective they are less inhibitory than the complete media and the colonies are so distinct as to be easily isolated. In this case, it is advisable to prepare 1:100 and 1:1000 dilutions of the plaque sample and spread 0·05 ml on each plate.

No selective medium has been devised for *L. aerogenes* although strains can sometimes be isolated from Omata and Disraely's medium. Colonies must usually be picked from blood agar plates and once again 1:100 and 1:1000 dilutions of plaque are convenient. Confusion of the colonies with those of fusobacteria is possible, but the latter are usually smoother, less rhizoid and have a "flecked" appearance. The production of large volumes of gas from carbohydrates by *L. aerogenes* clearly differentiates it from oral fusobacteria.

Both *Leptotrichia* spp grow well on blood agar, and strains can be maintained on this medium by serial transfer for many months. Subculture can still be made successfully after up to 10 days' incubation. For fluid culture, *L. buccalis* will grow well on the selective medium base, or almost any standard medium with the addition of 0·5% cysteine HCL with a paraffin seal. *L. aerogenes* is more exacting in its requirements and 1–2% horse or bovine serum should be added to good quality nutrient broth with 0·5% yeast extract (Difco).

Basal media for biochemical tests for *L. buccalis* have been described by Kasai (1965) and for *L. aerogenes* by Hofstad (1967a). Hofstad enriched his medium with ascitic fluid, but we have successfully replaced this with bovine serum (Oxoid).

Fusobacteria

Fusobacteria can readily be isolated from the mouth. Early classification of oral anaerobes as fusobacteria on the basis of their Gram-reaction and fusiform shape resulted in the description of several species. One of the most thorough studies on these organisms, that of Baird-Parker (1960), suggested that there are three groups and that groups 2 and 3 should probably be placed together. The groups were named as follows:

Group
1. *Fusobacterium nucleatum*
2. *Fusobacterium polymorphum*
3. *Fusiformis fusiformis*

Separation between *F. nucleatum* and *F. polymorphum* is based upon cell morphology and is still accepted (Burnett and Scherp, 1968). *F. plauti-vincente, F. fusiforme* and *F. dentium* are generally thought to be synonymous with *L. buccalis* (Baird-Parker, 1960; Omata and Braunberg, 1960; Morris, 1954).

Biochemical tests have not been successful for the differentiation of species, since most strains are similar in their reactions. Glucose, fructose, galactose and sometimes sucrose are weakly fermented, giving a final pH between 5·8–6·3. The fermentative activities of fusobacteria and leptotrichia have been studied by Jackins and Barker (1950) and amino acid metabolism of fusobacteria by Loesche and Gibbons (1968). All strains produce H_2S from media containing cysteine and are indole positive. Fusobacteria have been differentiated from other Gram-negative anaerobic rods by numerical taxonomic methods and shown to have a close relationship to *Sphaerophilus* (Barnes and Goldberg, 1968).

Serological differentiations of fusobacteria has been attempted, and cross-reaction of strains has been shown in precipitin tests (Araujo, Varah and Mergenhagen, 1963). Recent work by Kristoffersen (1969a, b) has demonstrated precipitin lines in gel diffusion studies which correspond to lipolysaccharide and "acid hapten" antigens, associated with the cell wall.

In ulcerative gingivitis, high numbers of fusobacteria can be found in saliva and gingival plaque (Hadi and Russell, 1968, 1969). However, the pathogenic role of these organisms in the mouth has not been established with any certainty.

The selective media of Baird-Parker (1957) and Omata and Disraely (1956) give good results for isolation or oral fusobacteria. Fairly distinct colonies are produced on blood agar; these are 1–3 mm in diam after 4 days' incubation, smooth translucent, greyish-white often with a characteristic "flecked" appearance and sometimes with a blue or yellowish tinge (Fig. 4). Microscopically the cells (Fig. 3) are Gram-negative 0·5–0·8 μ wide and either short (3–7 μ) or long (5–25 μ). The longer type may form filaments up to 100 μ in length. The short cell type may be called *F. nucleatum* and the presence of intracellular basophilic granules is said to be characteristic of this species. However, intracellular granules may also be demonstrated in the longer cell type *F. polymorphum*.

Fusobacteria can be isolated from salivary deposit or from plaque, especially when taken from the gingival area. Samples may be plated directly on the selective media, or 0·05 ml amounts of 1:50 and 1:100 dilutions of plaque suspension may be used.

Strains can be maintained by weekly serial subculture on blood agar. Freeze-drying is usually successful and strains have been recovered after storing for 3 years. Fluid cultures can be made in fairly large amounts

(5–10 litres) in the selective media bases, or in nutrient broth supplemented with yeast extract (0·5%), glucose (0·5%) and cysteine hydrochloride (0·2%). If the container is well filled and sealed with liquid paraffin before autoclaving, no further anaerobic methods are necessary. Some studies on the growth requirements of fusobacteria have been carried out (Omata, 1959; Coles, 1968).

Bacteroides

The principal habitat of the Gram-negative anaerobic rods in the mouth is the gingival sulcus and its associated gingival plaque. Studies on this area have shown that *Bacteroides* spp can represent a high proportion of the total flora. A simple scheme for the identification of oral Gram-negative anaerobic rods has been published by Loesche and Gibbons (1965).

B. *melanogenicus* has been the most intensively studied species. A second oral species, B. *oralis*, has been proposed by Loesche, Socransky and Gibbons (1964) and organisms resembling B. *corrodens* (Eiken, 1958) can be isolated from plaque samples and have also been isolated from the blood following tooth extraction (Khairat, 1966).

Animal experiments have shown that B. *melaninogenicus* is a necessary component if mixtures of oral organisms are to produce skin lesions (MacDonald *et al.*, 1960, 1968; MacDonald and Gibbons, 1962). B. *melaninogenicus* has been shown to increase in numbers in the gingival crevice with age, where it is usually absent up to the age of 5–6 years but almost universally present by adolescence (Bailit, Baldwin and Hunt, 1964; Kelstrup, 1966). These and other characteristics suggest a possible role for this species in the aetiology of periodontal disease.

B. *melaninogenicus* produces black colonies on blood agar after 4–7 days incubation, pigmentation is thought to be caused by haematin (Schwabacher, Lucas and Rimington, 1947). Recently, however, pigmentation in *Bacteroides* spp has been attributed to the production of extracellular ferrous sulphide (Tracy, 1969) Cells are Gram-negative 0·4–0·6 μ wide and varying in length from 2–10 μ. Oral strains are active against several carbohydrates; most ferment glucose, galactose and lactose, and produce indole and H_2S (Sawyer, MacDonald and Gibbons, 1962). Strains are actively proteolytic and produce a collagenase (Gibbons and MacDonald, 1961; Sawyer, MacDonald and Gibbons, 1962). Lipolysaccharide endotoxins have been extracted from oral strains (Mergenhagen, Hampp and Scherp, 1961; Hofstad, 1968, 1969*a*).

All strains of B. *melaninogenicus* have a requirement for haemin, while some also require menadione (Gibbons and MacDonald, 1960). Blood agar will support the initial growth of oral strains in samples and identification of the colonies is made relatively easy by their pigmentation. Blood agar supplemented with 5 μg/ml of menadione (added aseptically to the cooled base) provides a good medium for subculture and purification. Vancomycin blood agar (7·5 μg/ml vancomycin) can be used as a selective medium modified from that of McCarthy and Snyder (1963). Other selective media, such as those described for the isolation of intestinal *Bacteroides* spp, have not been applied to the study of the oral flora but would probably be effective. A basal medium for identification tests has been employed by Sawyer, MacDonald and Gibbons (1962). Hofstad (1968) used a medium supplemented with ascitic fluid for broth cultures prior to endotoxin extraction. Information on antibiotic sensitivities of *Bacteroides* and other Gram-negative anaerobic rods is contained in the paper of Finegold, Harada and Miller (1967).

Strains of *B. melaninogenicus* may be maintained by weekly subculture on blood agar containing menadione. Freeze-drying of strains is usually successful, especially if young colonies are taken before they become markedly pigmented. However, some strains do not survive freeze-drying.

Veillonella

Veillonella spp are amongst the most numerous anaerobic organisms in the mouth. The generally accepted description of the oral species is that of Rogosa (1964, 1965) and Rogosa and Bishop (1964a, b). They are regular Gram-negative cocci, 0·3–0·5 μ in diam, usually arranged in pairs or masses. They are non-motile, non-sporing and strictly anaerobic, often with a requirement for CO_2.

These organisms do not ferment carbohydrates, but several organic acids, including lactic, succinic, pyruvic and fumaric acids, are metabolized by resting cells to produce CO_2 and H_2. Lactic acid is attacked weakly by resting cells, but in growing cultures it is metabolized to propionic and acetic acids with CO_2 and H_2 release. A recent study (Kafkewitz and Delwiche, 1969) has shown that *Veillonella* assimilates C^{14} ribose which can be detected in the nucleic acids of the cells. These authors also confirm the absence of hexokinase in *Veillonella* (Rogosa, Krichevsky and Bishop, 1965) and note that phosphoglyceromutase and pyruvate kinase could not be detected in a strain of *V. alkalescens*. All strains are indole-negative, reduce nitrate, and will produce H_2S in the presence of cysteine.

The two species found in the human mouth are *V. parvula* and *V. alkalescens*. Separation of these species can be made upon their requirement for putrescine and cadaverine, using the medium described by Rogosa and Bishop (1964a, b). Generally human *V. alkalescens* strains require these compounds whilst human *V.parvula* strains do not. Further details of human and animal *Veillonella* spp are given by Rogosa (1965).

Serological groupings have also been recorded (Rogosa, 1965) eight groups being proposed on the basis of whole cell agglutination tests. A major antigenic component is associated with a lipopolysaccharide endo-toxin (Mergenhagen and Varah, 1963; Mergenhagen *et al.*, 1962; Bladen and Mergenhagen, 1964). Sims (1960) has published work on the charac-teristics of some oral strains of *Veillonella* and their relationship to dental caries.

Veillonella can be isolated from most sites in the mouth, especially plaque and saliva. The media described by Rogosa (1956), Rogosa, Fitz-gerald, MackIntosh and Beaman (1958) give excellent results and we routinely use the earlier formula with streptomycin. Plates should be incubated in 95% H_2 + 5% CO_2 for 4 days. Volumes of 0·05 ml of 1:10 and 1:100 dilutions of plaque samples on selective plates usually give satisfactory results in quantitative studies. These organisms will grow on blood agar, and also in the selective medium base without agar, dye or antibiotic.

Oral Spirochaetes

Spirochaetes are common in the mouth, particularly in the gingival and inter-dental areas (Fig. 5). In patients with periodontal disease they are frequently found in high numbers in association with fusobacteria and other anaerobes. Numbers are low in pre-dentate infants and edentulous adults, the presence of teeth apparently favouring their growth. Soft tissue inflammation, especially of the gingivae, is usually accompanied by an increase in the spirochaetal flora.

The nomenclature and taxonomy of the oral spirochaetes is confused. Bergey's Manual (7th Edition, 1957) lists four species: *Borrelia buccalis*, *Borrelia vincentii*, *Treponema microdentium* and *Treponema mucosa*. Many more species have been reviewed by Roseburky (1962). Much of the work on oral spirochaetes during the period 1940–1960 was done by Dr E. G. Hampp and his associates at the National Institute of Dental Research, Bethesda (Omata and Hampp, 1961). A more recent electron microscope study by Listgarten and Socransky (1965) suggest that oral spirochaetes should be divided into three main groups based on morphological charac-teristics.

Group 1. "Small" spirochaetes. These have a protoplasmic cylinder of 0·1–0·25 μ diam, an easily disrupted outer envelope consisting of polygonal structural subunits, and either one or two axial fibrils arising at each end.

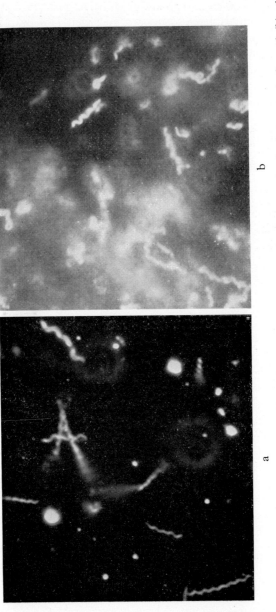

b

a

FIG. 5. Oral spirochaetes. (a), dark field (× 2000), wet preparation from gingival area in the human mouth, and (b), dark field (× 2000), wet preparation of agar taken from the outer haze of a Petri dish membrane culture.

FIG. 5. Oral spirochaetes. (c), Petri dish membrane culture of gingival sample. Note the two outer haze rings of growth. The first, marked by the line of dashes on the base of the plate was evident after 7 days' incubation, the second developed after incubation for a further 7 days.

These axial fibrils give rise to the "1–2–1" and "2–4–2" types of arrangements that are described. Recently, Socransky *et al.* (1969) have described three separate species of small spirochaetes, *T. denticola, T. macrodentium* and *T. oralis* (proposed new species).

Group 2. "Intermediate" spirochaetes. Here the protoplasmic cylinder is 0·2–0·5 μ in diam and there are 3–20 axial fibrils. This group includes *B. vincentii.*

Group 3. "Large" spirochaetes. The cytoplasmic cylinder is 0·5 μ or more in diam, with 12–20 axial fibrils. The species *Borrelia buccalis* described by Hampp (1954) is included in this group.

As Listgarten and Socransky (1965) point out, the classification of spirochaetes on morphological grounds alone is unsatisfactory. Improved cultivation methods together with further biochemical and serological studies should throw more light on these neglected micro-organisms.

P

Isolation methods

Although the isolation and cultivation of oral spirochaetes is possible, it is not routinely done by most laboratories. The presence of spirochaetes in oral samples is commonly diagnosed by dark-field microscopy alone. In fact, according to Bergeron (1883) certain oral spirochaetes were successfully cultivated by Pasteur and Netter.

All isolation methods depend basically on the ability of motile spirochaetes to penetrate culture media and migrate away from non-motile oral contaminants. The principles of isolation were advocated by Noguchi (1912) and three variations are currently used:

(1) Stab culture in tubes (original Noguchi technique).
(2) Stab into wells cut in deep agar media.
(3) Inoculation on to membrane filter on surface of solid medium.

In each case the growth of spirochaetes is recognized by the appearance of a haze within the gel (Fig. 5) which may then be subcultured for purification.

The membrane filter method of isolation is probably the most useful and will be described briefly below. All isolation media that have been described are highly enriched with serum, blood or ascitic fluid, and contain agar. Blake (1968) described the following medium:

Ion agar	0·8%	
Tryptone	3·0%	
Yeast extract	0·5%	w/v
Sodium chloride	0.25%	
L-cysteine hydrochloride	0·075%	
Glucose	0·5%	
Heat-inactivated horse serum	10·0%	v/v

pH adjusted to 7·4, sterilized at 121°/15 min and serum added when medium had cooled.

A medium of similar composition is employed in the V.D. Reference Laboratory P.H.L.S., London Hospital (Dr N. Perera, personal communication). This consists of a 50:50 mixture of Rajkovic's medium (Rajkovic, 1966) and Brain Heart Infusion Broth (Oxoid), to which is added 10% calf serum and 0·7% ion agar. Davies (1969) has used Spirolate medium (B.B.L.) supplemented with serum and agar, and several other suitable media are described by Socransky *et al.* (1969).

Sterile Millipore filters (Millipore (U.K.) Ltd.) are laid upon the surface of plates of the medium. Material from the mouth is placed on the surface of the filter and incubated anaerobically at 35–37° for several days. Spirochaetes in the inoculum pass through the filter and grow into the medium, whilst other organisms are confined to the surface. It is important that the plates should be reasonably dry and that the inoculum should be in a small volume of liquid to avoid surface spread of contaminants.

Blake (1968) found that $0.22\,\mu$ filters gave the best results although some workers favour $0.1\,\mu$ ones. He also described a modified technique using pairs of different diameter filters cemented to Perspex rings which apparently improved the isolation. For examination and subculture the filters are removed and small pieces of agar containing the haze of growth are cut out. Subcultures from primary isolates have been successful using shake culture or pour-plate techniques. A liquid medium without agar was used by Blake (1968), and up to five successive liquid to liquid subcultures were achieved. Granular and cystic stages are often found in cultures, but their significance is poorly understood.

Comparatively few studies have been made of the metabolic, serological and other properties or oral spirochaetes. Much of the published information available is summarized in the relevant chapters of the books by Rosebury (1962) and Burnett and Scherp (1968). A recent paper by Socransky et al. (1969) gives some useful information on the characteristics of small spirochaetes.

Other Anaerobic Microorganisms in the Mouth

There are several other types of anaerobe found in the mouth whose identity and classification are poorly understood. These include anaerobic streptococci, diphtheroids and a variety of Gram-negative motile rods. Some of these are considered in this section.

Anaerobic Streptococci (Peptostreptococcus)

Anaerobic streptococci have been isolated from the mouth, where they may constitute from 4–13% of the viable count from certain sites (Burnett and Scherp, 1968). Several species have been described but their classification is inadequate. Bahn, Kung and Hayashi (1966) have studied a number of oral isolates by cell wall analysis and serological techniques. These isolates were obtained from cases of osteomyelitis of the mandible and deep submandibular abscesses.

There appear to be no generally accepted special media or techniques for isolating oral anaerobic streptococci. They will grow on blood agar in 95% H_2 + 5% CO_2 but it is difficult to distinguish them from other facultative streptococci which grow under these conditions. Thomas and Hare (1954) found that the addition of 0.01% sodium oleate to culture media enhanced the growth of Groups III and IV. These workers suggested the criterion of three successive failures to grow aerobically to distinguish anaerobic strains from facultative streptococci should be used. However, it must be remembered that some oral streptococci are CO_2 dependent.

Anaerobic Diphtheroids

There is considerable confusion surrounding the classification of both facultative and strictly anaerobic diphtheroid organisms. In practice, almost any Gram-positive, pleomorphic, non-motile rod that is not obviously a lactobacillus or actinomyces is placed in the "diphtheroid bag". In a study of 50 anaerobic diphtheroids from the mouth (Rasmussen et al., 1966), 7 strains were identified as *Propionibacterium acnes*, 6 were catalase positive corynebacteria (unspecified) and the remainder were assigned to the genus *Actinomyces*.

Isolation of this ill-defined group of organisms may be achieved using horse blood agar incubated for 3–10 days at 37° in 95% + 5% CO_2. Subcultures of strains with the appropriate microscopic appearance may be maintained on blood agar or in thioglycollate broth.

Oral Gram-negative motile anaerobic rods

The classification of organisms in this general category is particularly confused. Three groups were proposed by MacDonald (1953). A scheme for their identification has been described by Loesche and Gibbons (1965).

Vibrio sputorum

This species has been studied by Loesche et al. (1965). It is a Gram-negative, straight or slightly curved rod, 0·4–0·8 μ in size. Motility is usually only present in young cultures. When flagella are present they are terminal or subterminal.

Isolation

Growth in all media is enhanced by the addition of 0·1% KNO_3. Loesche and co-workers (1965) recommend an initial isolation procedure similar to that described for spirochaetes, using 0·45 or 0·3 μ millipore filters on the surface of an enriched agar plate. Growth in the agar is subcultured into a basal medium consisting of:

> Thioglycollate Broth, without Dextrose, (BBL)
> Yeast Extract (Difco), 0·2%
> KNO_3, 0·1%

Isolated strains may be maintained on blood agar plates incubated in 95% H_2 + 5% CO_2. Grey colonies, round with thin irregular edge, 1–2 mm in diam, shiny, low convex smooth are formed, sometimes with a-haemolysis.

Strictly speaking, *V. sputorum* is micro-aerophilic and will grow in the presence of 5% O_2. The oral vibrios have similarities to *V. bubulus* and *V. foetus* and all three differ in many respects from the true vibrios. It has been proposed that they should be included in the genus *Campylobacter* (for additional details, this volume p. 212).

Selenomonas sputigenum (Spirillum sputigenum)

There is considerable confusion in the literature concerning the identity of this organism which was called *Spirullum sputigen* by MacDonald (1953) in his classification of oral motile nonsporulating anaerobic rods. There is some doubt as to whether the organism is a bacterium (Loesche and Gibbons, 1965) or a protozoan (Bisset and Davis, 1960). A recent report describes the appearance of the flagella which arise from the concave side of the curved organism (Jeynes and Bisset, 1968).

The organisms are strictly anaerobic and may be grown on solid or semi-solid media containing glucose and either blood or serum. MacDonald and Madlener (1957) described a veal-infusion medium with serum and a surface active agent. On blood agar, smooth convex greyish-yellow colonies less than 0·5 mm in diam have been described (Burnett and Scherp, 1968).

Bisset and Davis (1960) state that immediate inoculation of samples is essential and suggest that they should be emulsified in a drop of serum and incubated in 90% H_2 + 10% CO_2 at 37° for 3–4 days. Subcultures in Brewer's medium or fusobacterium medium (without dye or antibiotics) may be made. The organisms tend to grow more readily in mixed culture, especially with fusobacteria.

References

AJELLO, L., GEORG, L. K., KAPLAN, W. & KAUFMAN, L. (1963). C.D.C. Laboratory Manual for Medical Mycology. *P.H.S. Pub. N.* 994, U.S. Govt. Printing Office.

ARAUJO, W. C. DE, VARAH, E. & MERGENHAGEN, S. E. (1963). Immunochemical analysis of human oral strains of *Fusobacteria* and *Leptotrichia*. *J. Bact.*, **86**, 837.

BABOOLAL, R. (1968). Identification of filamentous microorganisms of human dental plaque by immunofluorescence. *Caries Res.*, **2**, 273.

BABOOLAL, R. (1969). Cell wall analysis of oral filamentous bacteria. *J. gen. Microbiol.*, **58**, 217.

BAHN, A. N., KUNG, P. C. Y. & HAYASHI, J. A. (1966). Chemical composition and serological analysis of the cell wall of *Peptostreptoccus*. *J. Bact.*, **91**, 1672.

BAILIT, H. L. BALDWIN, D. C. & HUNT, E. E. JR. (1964). The increasing prevalence of gingival *B. melaninogenicus* with age in children. *Archs oral Biol.*, **9**, 435.

BAIRD-PARKER, A. C. (1957). Isolation of *Leptotrichia buccalis* and *Fusobacterium* species from oral material. *Nature, Lond.*, **180**, 1056.

BAIRD-PARKER, A. C. (1960). The classification of fusobacteria from the human mouth. *J. gen. Microbiol.*, **22**, 458.

BAIRD-PARKER, A. C. & DAVIS, G. H. G. (1958). The morphology of the *Leptotrichia* species. *J. gen. Microbiol.*, **19**, 446.

BARNES, E. M. & GOLDBERG, H. S. (1968). The relationships of bacteria within the Family Bacteroidaceae as shown by numerical taxonomy. *J. gen. Microbiol.*, **51**, 313.

BATTY, I. (1958). *Actinomyces odontolyticus*, a new species of actinomycete regularly isolated from deep carious dentine. *J. Path. Bact.*, **75**, 455.

BERGERON, J. (1883). Stomatites. Dictionnaire Encyclopédique des Sciences Médicales, Ste Sue, 91 (U51 12 of 3rd ed.), Paris.

BISSET, K. A. & DAVIS, G. H. G. (1960). *The Microbial Flora of the Mouth.* London: Heywood and Company Ltd.

BLADEN, H. A. & MERGENHAGEN, S. E. (1964). Ultrastructure of *Veillonella* and morphological correlation of an outer membrane with particles associated with endotoxic activity. *J. Bact.*, **88**, 1482.

BLAKE, G. C. (1968). The microbiology of acute ulcerative Gingivitis with reference to the culture of oral trichonomads and spirochaetes. *Proc. R. Soc. Med.*, **61**, 131.

BLANK, C. H. & GEORG, L. K. (1968). The use of fluorescent antibody methods for the detection and identification of *Actinomyces* species in clinical material. *J. Lab. Clin. Med.*, **71**, 283.

BREED, R. S., MURRAY, E. G. D. & SMITH, N. R. (1957). Bergey's Manual of Determinative Bacteriology, 7th ed. London: Baillière, Tindall & Cox.

BOONE, C. J. & PINE, L. (1968). Rapid method for characterization of *Actinomycetes* by cell wall composition. *Appl. Microbiol.*, **16**, 279.

BROCK, D. W. & GEORG, L. K. (1969). Characterization of *Actinomyces israeli* Serotypes 1 and 2. *J. Bact.*, **97**, 589.

BUCHANAN, B. B. & PINE, L. (1962). Characterization of a propionic acid producing actinomycete. *Actinomyces propionicus*, sp. nov. *J. gen. Microbiol.*, **28**, 305.

BURNETT, G. W. & SCHERP, H. W. (1968). *Oral Microbiology and Infectious Disease.* 3rd Edition. Baltimore: The Williams and Wilkins Company.

COLES, R. S. JR. (1968). Some growth factors for *Fusobacterium polymorphum*. *J. Bact.*, **96**, 2183.

CUMMINS, C. S. (1962). Chemical composition and antigenic structure of cell walls of *Corynebacterium, Mycobacterium, Nocardia, Actinomyces* and *Arthrobacter. J. gen. Microbiol.*, **28**, 35.

CUMMINS, C. S. & HARRIS, H. (1958). Studies on cell wall composition and taxonomy of Actinomycetales and related groups. *J. gen. Microbiol.*, **18**, 173.

DAVIES, R. M. (1969). The *in vitro* sensitivity of oral spirochaetes to Metronidazole Abstract. British Division, I.A.D.R. *J. Dent. Res.,* **48.** *Supplement to No.* 6, 1101.

DAVIS, G. H. G. & BAIRD-PARKER, A. C. (1959a). Cell wall composition of *Leptotrichia* Sp. *Nature, Lond.,* **183,** 1206.

DAVIS, G. H. G. & BAIRD-PARKER, A. C. (1956b). The classification of certain filamentous bacteria with respect to their chemical composition. *J. gen. Microbiol.,* **21,** 612.

DAVIS, G. H. G. & FREER, J. H. (1960). Studies upon an oral aerobic actinomycete. *J. gen. Microbiol.,* **23,** 163.

DE WEESE, M. S. GERENCSER, M. A. & SLACK, J. M. (1968). Quantitative analysis of *Actinomyces* cell walls. *Appl. Microbiol.,* **16,** 1713.

EIKEN, M. (1958). Studies on an anaerobic rod-shaped Gram-negative microorganism. *Bacteroides corrodens.* n. sp. *Acta. path. microbiol. scand.,* **43,** 404.

ERICKSON, D. (1940). Pathogenic anaerobic organisms of the *Actinomyces* group. *Br. med. Res. Council Spec. Rept. Ser.,* No. 240. pp. 1–63.

FINEGOLD, S. M., HARADA, N. E. & MILLER, L. G. (1967). Antibiotic susceptibility patterns as aids in classification and characterisation of Gram-negative anaerobic bacilli. *J. Bact.,* **94,** 1443.

GEORG, L. K. & BROWN, J. M. (1967). *Rothia,* gen. nov. An aerobic genus of the family *Actinomycetaceae. Int. J. Syst. Bacteriol.,* **17,** 79.

GEORG, L. K., PINE, L. & GERENCSER, M. A. (1969). *Actinomyces viscosus,* comb. nov. a catalase positive, facultative member of the genus *Actinomyces. Int. J. Syst. Bacteriol.,* **19,** 291.

GEORG, L. K., ROBERTSTADT, G. W. & BRINKMAN, S. A. (1964). Identification of species of *Actinomyces, J. Bact.,* **88,** 477.

GEORG, L. K., ROBERTSTADT, G. W., BRINKMAN, S. A. & HICKLIN, M. D. (1965). A new pathogenic anaerobic *Actinomyces* species. *J. infect. Dis.* **115,** 88.

GERENCSER, M. A., LANDFRIED, S. & SLACK, J. M. (1969). Fluorescent antibody and cultural studies of Actinomycetes in dental calculus. *Abstract. No.* 284. 47th. General Meeting, I.A.D.R. North American Division.

GERENCSER, M. A. & SLACK, J. M. (1967). Isolation and characterization of *Actinomyces propionicus. J. Bact.,* **94,** 109.

GERENCSER, M. A. & SLACK, J. M. (1969). Identification of human strains of *Actinomyces viscosus. Appl. Microbiol.,* **18,** 80.

GIBBONS, R. J. & MACDONALD, J. B. (1960). Hemin and vitamin K compounds as required factors for the cultivation of *B. melaninogenicus. J. Bact.,* **80,** 164.

GIBBONS, R. J. & MACDONALD, J. B. (1961). Degradation of collagenous substrates by *B. melaninogenicus. J. Bact.,* **81,** 614.

GILMOUR, M. N., HOWELL JR. A. & BIBBY, B. G. (1961). The classification of organisms termed *Leptotrichia* (*Leptothrix*) *buccalis.* 1. Review of the literature and proposed separation into *Leptotrichia buccalis* Trevisan, 1879 and *Bacterionema* gen. nov., *B. matruchotii* (Mendel 1919.) comb. nov. *Bact. Rev.,* **25,** 131.

GUSTAFSON, K. L., KROEGER, A. V. & VAICHULIS, E. M. K. (1966). Chemical characteristics of *Leptotrichia buccalis* endotoxin. *Nature, Lond.,* **212,** 301.

HADI, A. W. & RUSSELL, C. (1968). Quantitative estimations of fusiforms in saliva from normal individuals and cases of acute ulcerative gingivitis. *Archs oral Biol.*, **13**, 1371.

HADI, A. W. & RUSSELL, C. (1969). Fusiforms in gingival material. *Br. dent. J.*, **126**, 82.

HAMILTON, R. D. & ZAHLER, S. A. (1957). A study of *Leptotrichia buccalis*. *J. Bact.*, **73**, 386.

HAMPP, E. G. (1947). Bacteriologic investigations of the oral spirochaetal flora in ulcerative stomatitis (Vincents Infection). *Am. J. Orthod. Oral Surg.*, **33**, 666.

HAMPP, E. G. (1954). *Borrelia buccalis* isolation, pure cultivation and morphologic characteristics. *J. dent. Res.*, **33**, 660.

HARZ, C. O. (1878). *Actinomyces bovis*, ein neuer Schimmel in den Geweben des Rindes. *Deut. Ztschr. Thiermed.*, **5**, 125.

HOFSTAD, Y. (1967a). An anaerobic filamentous organism possibly related to *L. buccalis*. 1. *Acta. path. microbiol. scand.*, **69**, 543.

HOFSTAD, T. (1967b). An anaerobic filamentous organism possibly related to *L. buccalis*. 2. *Acta. path. Microbiol. scand.*, **70**, 461.

HOFSTAD, T. (1968). Chemical characteristics of *B. melaninogenicus* endotoxin. *Archs oral. Biol.*, **13**, 1149.

HOFSTAD, T. (1969a). Serological properties of lipopolysaccharide from oral strains of *B. melaninogenicus*. *J. Bact.*, **97**, 1078.

HOFSTAD, T. (1969b). Ultrastructure of an anaerobic filamentous oral microorganism *J. gen. Microbiol.*, **57**, 221.

HOFSTAD, T. & SELVIG, K. A. (1969). Ultrastructure of *L. buccalis*. *J. gen. Microbiol.*, **56**, 23.

HOWELL, A. JR. (1963). A filamentous microorganism isolated from periodontal plaque in hamsters. *Sabouraudia.*, **3**, 81.

HOWELL, A. JNR., JORDAN, H. V., GEORG, L. K. & PINE, L. (1965). *Odontomyces viscosus*, gen. nov. spec. nov. a filamentous microorganism isolated from periodontal plaque in hamsters. *Sabouraudia.*, **4**, 65.

HOWELL, A. JNR., MURPHY, W. C., PAUL, F. & STEPHAN, R. M. (1959). Oral strains of Actinomyces. *J. Bact.*, **78**, 82.

HOWELL, A., JNR., STEPHAN, R. M. & PAUL, F. (1962). Prevalence of *A, israeli, A. naeslundi, B. matruchottii* and *C. albicans* in selected areas or the oral cavity and saliva. *J. dent. Res.*, **41**, 1050.

JACKINS, H. C. & BARKER, H. A. (1950). Fermentative processes of the fusiform bacteria. *J. Bact.*, **61**, 101.

JEYNES, M. H. & BISSET, K. A. (1968). The flagella of *Selenomonas sputigena* in relation to cell wall growth and nuclear division of microorganisms. *Giornale di Microbiologia*, **16**, 65.

KAFKEWITZ, D. & DELWICHE, E. A. (1969). Utilization of D-ribose by *Veillonella*. *J. Bact.*, **98**, 903.

KASAI, G. J. (1961). A study of *Leptotrichia buccalis*. 1. *J. dent. Res.*, **40**, 800.

KASAI, G. J. (1965). A study of *Leptotrichia buccalis*. 2. *J. dent. Res.*, **44**, 1015.

KELSTRUP, J. (1966). The incidence of *B. melaninogenicus* in human gingival sulci, and its prevalence in the oral cavity at different ages. *Periodontics.*, **4**, 14.

KHAIRAT, O. (1966). An effective antibiotic cover for the prevention of endocarditis following dental and other post-operative bacteraemias. *J. clin. Path.*, **19**, 561.

KING, S. & MEYER, E. (1963). Gel diffusion technique in antigen—antibody reactions of *Actinomyces* species and "anaerobic diptheroids". *J. Bact.*, **85**, 186.

KRISTOFFERSEN, T. (1969a). Immunochemical studies of oral fusobacteria. 1. *Acta. path. microbiol. scand.*, **77**, 235.

KRISTOFFERSEN, T. (1969b). Immunochemical studies of oral fusobacteria. 2. *Acta. path. microbiol. scand.*, **77**, 247.

LAMBERT, F. W. JR., BROWN, J. M. & GEORG, L. K. (1967). Identification of *Actinomyces israeli* and *Actinomyces naeslundi* by fluorescent antibody and agar-gel diffusion techniques. *J. Bact.*, **94**, 1287.

LISTGARTEN, M. A. & SOCRANSKY, S. S. (1965). Electron microscopy as an aid in the taxonomic differentiation of oral spirochaetes. *Archs oral Biol.*, **10**, 127.

LOESCHE, W. J. & GIBBONS, R. J. (1965). A practical scheme for identification of the most numerous oral Gram-negative anaerobic rods. *Archs oral Biol.*, **10**, 723.

LOESCHE, W. J., GIBBONS, R. J. & SOCRANSKY, S. S. (1965). Biochemical characteristics of *Vibrio sputorum* and relationship to *Vibrio bubulus* and *Vibrio fetus*. *J. Bact.*, **89**, 1109.

LOESCHE, W. J. & GIBBONS, R. J. (1968). Amino acid fermentation by *Fusobacterium nucleatum*. *Archs oral Biol.*, **13**, 191.

LOESCHE, W. J., SOCRANSKY, S. S. & GIBBONS, R. J. (1964). *Bacteroides oralis* proposed new species isolated from the oral cavity of man. *J. Bact.*, **88**, 1329.

MACDONALD, J. B. (1953). The motile nonsporulating anaerobic rods of the oral cavity. Thesis, University of Toronto.

MACDONALD, J. B. & GIBBONS, R. J. (1962). The relationship of indigenous bacteria to periodontal disease. *J. dent. Res.*, (Suppl. to No. 1.) **41**, 320.

MACDONALD, J. B., GIBBONS, R. J. & SOCRANSKY, S. S. (1960). Metabolism of oral tissues. Bacterial mechanisms in periodontal disease. *Ann. N. Y. Acad. Sci.*, **85**, 467.

MACDONALD, J. B. & MADLENER, E. M. (1957). Studies on the isolation of *Spirillum sputigenum*. *Canad. J. Microbiol.*, **3**, 679.

MACDONALD, J. B., SOCRANSKY, S. S. & GIBBONS, R. J. (1968). Aspects of the pathogenesis of mixed anaerobic infections of mucous membranes. *J. dent. Res.*, **42**, 529.

MCCARTHY, C. & SNYDER, M. L. (1963). Selective medium for Fusobacteria and Leptotrichia. *J. Bact.*, **86**, 158.

MELVILLE, T. H. (1965). A Study of the overall similarity of certain Acitinomycetes, mainly of oral origin. *J. gen. Microbiol.*, **40**, 309.

MERGENHAGEN, S. E., HAMPP, E. G. & SCHERP, H. W. (1961). Preparation and biological activities of endotoxins from oral bacteria. *J. infect. Dis.*, **108**, 304.

MERGENHAGEN, S. E. & VARAH, E. (1963). Serologically specific lipolysaccharides from oral *Veillonella*. *Arch. oral Biol.*, **8**, 31.

MERGENHAGEN, S. E., ZIPKIN, I. & VARAH, E. (1962). Immunological and chemical studies on oral *Veillonella* endotoxin. *J. Immun.*, **88**, 482.

MORRIS, E. O. (1954). The bacteriology of the oral cavity. 6. *Brit. dent. J.*, **XCVII**, 283.

NOGUCHI, H. (1912). Cultural studies of mouth spirochaetes (*Trepenonema microdentium* and *macrodentium*). *J. exp. Med.*, **15**, 81.

OMATA, R. R. (1959). Studies upon the nutritional requirements of the fusobacteria. *J. Bact.*, **96**, 2183.

OMATA, R. R. & BRAUNBERG, R. C. (1960). Oral fusobacteria. *J. Bact.*, **80**, 737.

OMATA, R. R. & DISRAELY, M. N. (1956). A selective medium for oral fusobacteria. *J. Bact.*, **72**, 677.

OMATA, R. R. & HAMPP, E. G. (1961). Proteolytic activities of some oral spirochaetes. *J. dent. Res.*, **40**, 171.

PINE, L. & BOONE, C. J. (1967). Comparative cell wall analyses of morphological forms within the genus *Actinomyces*. *J. Bact.*, **94**, 875.

PINE, L. & GEORG, L. K. *Int. J. Syst. Bacteriol.* (In press). See Gerencser and Slack (1969).

RAJKOVIC, A. D. (1966). Beef serum supplement to media for the cultivation of the Noguchi strain of *T. pallidum*. *Zeit, Med. Mikrobiol. v. Immunol.*, **152**, 100.

RASMUSSEN, E. G., GIBBONS, R. J. & SOCRANSKY, S. S. (1966). A taxonomic study of fifty Gram-positive anaerobic diphtheroids isolated from the oral cavity of man. *Arch. oral Biol.*, **11**, 573.

RITZ, H. L. (1967). Microbial population shifts in developing human dental plaque. *Arch. oral Biol.*, **12**, 1561.

ROGOSA, M. (1956). A selective medium for the isolation and enumeration of the *Veillonella* from the oral cavity. *J. Bact.*, **72**, 533.

ROGOSA, M. (1964). The genus *Veillonella*. 1. General cultural, ecological and biochemical considerations. *J. Bact.*, **87**, 162.

ROGOSA, M. (1965). The genus *Veillonella*. 4. Serological grouping, and genus and species emendations. *J. Bact.*, **90**, 704.

ROGOSA, M. & BISHOP, F. S. (1964a). The genus *Veillonella*. 2. Nutritional studies. *J. Bact.*, **87**, 574.

ROGOSA, M. & BISHOP, F. S. (1964b). The genus *Veillonella*. 3. Hydrogen sulphide production by growing cultures. *J. Bact.*, **88**, 37.

ROGOSA, M. R., FITZGERALD, J., MACKINTOSH, M. E. & BEAMAN, A. J. (1958). Improved medium for selective isolation of *Veillonella*. *J. Bact.*, **76**, 455.

ROGOSA, M., KRICHEVSKY, M. I. & BISHOP, F. S. (1965). Truncated glycolytic system in *Veillonella*. *J. Bact.*, **90**, 164.

ROSEBURY, T. (1962). *Micro-organisms Indigenous to Man*. McGraw-Hill Book Company, Inc.

ROTH, G. D. & THURN, A. N. (1962). Continued study of oral Nocardia. *J. dent. Res.*, **41**, 1279.

SAWYER, S. J., MACDONALD, J. B. & GIBBONS, R. J. (1962). Biochemical characteristics of *B. melaninogenicus*. *Arch. oral Biol.*, **7**, 685.

SCHWABACHER, H., LUCAS, D. R. & RIMINGTON, C. (1947). *Bacterium melaninogenicum*—A misnomer. *J. gen. Microbiol.*, **1**, 109.

SIMS, W. (1960). The characteristics of some oral strains of *Veillonella*. *Brit. dent. J.*, **108**, 73.

SLACK, G. L. & BOWDEN, G. H. (1965). Preliminary studies of experimental dental plaque *in vivo*. *Advances in Fluorine Research and Dental Caries Prevention*. **3**, 193. (Pergamon Press)

SLACK, J. M., LANDFRIED, S. & GERENCSER, M. A. (1969). Morphological, biochemical and serological studies on 64 strains of *Actinomyces israelii*. *J. Bact.*, **97**, 873.

SNYDER, M. L., BULLOCK, W. W. & PARKER, R. B. (1967). Morphology of Gram-positive filamentous bacteria identified in dental plaque by fluoresent antibody technique. *Arch. oral Biol.*, **12,** 1269.

SOCRANSKY, S. S., LISTGARTEN, M., HUBERSAK, C., COTMORE, J. & CLARK, A. (1969). Morphological and biochemical differentiation of three types of small oral spirochaetes. *J. Bact.*, **98,** 878.

THEILADE, E. & GILMOUR, M. N. (1961). An anaerobic filamentous organism. *J. Bact.*, **81,** 661.

THOMAS, C. G. A. & HARE, R. (1954). The classification of anaerobic streptococci and their isolation in normal human beings and pathological processes. *J. clin. Path.*, **7,** 300.

THOMPSON, L. & LOVESTEDT, S. A. (1951). An actinomyces like organism obtained from the human mouth. *Proc. Mayo. Clinic,* **26,** 169.

TRACY, O. (1969). Pigment production in *Bacteroides*. *J. Med. Microbiol.*, **2,** 309.

The Isolation of Microaerophilic Vibrios

J. A. MORRIS AND R. W. A. PARK

Microbiology Department, Reading University, Reading, England

Several vibrios exist which are generally unable to grow aerobically or anaerobically but which grow in an atmosphere of from 10 to 30% (v/v) CO_2 in air. Such organisms may be overlooked during routine isolation procedures unless suitable gaseous conditions are provided.

The microaerophilic vibrios, although morphologically similar to the type species of the genus *Vibrio, V. cholerae* (Park, 1961), have been shown to differ from it in DNA base ratios, and in several other respects, and so have been placed in a separate genus, *Campylobacter* (Sebald and Veron, 1963). The most widely known microaerophilic vibrio is *Campylobacter fetus* (i.e. *Vibrio fetus*; Smith and Taylor, 1919). Florent (1959) subdivided this species into two varieties, *venerealis* and *intestinalis*, and many workers have accepted this subdivision. However, the distinction between these two varieties is not so clear as was first thought (Park, Munro, Melrose and Stewart, 1962; Florent, 1963) and differentiation of types within the species remains difficult (but see also King, 1962). The only species besides *C. fetus* which was included in *Campylobacter* by Sebald and Veron (1963) was *C. bubulus* (i.e. *V. bubulus*; Florent, 1953). Loesche, Gibbons and Socransky (1965) showed that this organism is very similar to *V. sputorum* (Prevot, 1940) and they proposed that both types should be included in a single species. Other organisms which from the literature appear to be very similar to *C. fetus* and *C. bubulus* include *V. jejuni* (Jones, Orcutt and Little, 1931), *V. coli* (Doyle, 1948) and *V. faecalis* (Firehammer, 1965). We are unable to express an opinion about the justification of species rank for these three organisms because we have not examined any authentic strains.

Habitats of Microaerophilic Vibrios

Microaerophilic vibrios have been isolated by various workers from several animal species including man. Many isolations were from diseased material but some microaerophilic vibrios can be isolated regularly from animals showing no signs of disease. Some of the sites from which microaerophilic vibrios have been isolated are given in Table 1.

TABLE 1. Sites from which microaerophilic vibrios have been isolated

Animal*	Site	Effect
Cattle	Genital tract	Infertility in females or no clinical symptoms. No clinical symptoms in infected bulls.
	Gut and faeces	Diarrhoea or no clinical symptoms.
	Aborted foetuses	Abortion.
Sheep	Gut and faeces	No clinical symptoms.
	Aborted foetuses	Abortion.
Pigs	Gut and faeces	Diarrhoea, chiefly after weaning, or no clinical symptoms.
Man	Various	Abortion, fever, diarrhoea or no clinical symptoms.

* Microaerophilic vibrios also have been isolated from goats, antelopes, chickens, and carrion crows.

General Growth Conditions

Media

Information about three of the media used for routine cultivation of microaerophilic vibrios is given below but for chemically defined media see Smibert (1963) and Basden, Tourtellotte, Plastridge and Tucker (1968).

Thio blood agar (TBA)

For cultivation on a solid medium TBA is used (Morgan, 1957). In our laboratory the thio agar base is prepared from (%, w/v, in demineralized water): peptone (Evans), 0·5; Lab Lemco, 0·5; NaCl, 0·5; sodium thioglycollate, 0·1; Agar No. 3 (Oxoid), 1·5. The medium is adjusted to pH 7·0 and autoclaved (121° for 20 min). To prepare the complete medium the base is melted, cooled to 60° and then oxalated horse blood is added to give a final concentration of 10% (v/v). This mixture is kept at 60° for 10 min and then is poured into 90 mm Petri dishes, c. 15 ml being used per dish. Plates are dried at 55° and used if possible on the day of preparation. When correctly prepared the medium is reddish-brown.

Brewer's medium

The Oxoid dehydrated "Thioglycollate Medium (Brewer)" is reconstituted and dispensed in 10 ml amounts in 1 oz McCartney bottles. After autoclaving the bottles are not opened until they are inoculated. One of the constituents of the medium as supplied is methylene blue, which is an oxidation—reduction indicator. A medium which is the colour of nutrient broth or of which only the top few millimetres is green is suitable for inoculation. A medium which is green throughout should be held at boiling point for 10 min (when the dye will be reduced), rapidly cooled and then used immediately. A medium will not be suitable for vibrios if it has been reheated more than once.

Albimi brucella broth (ABB)

Brucella broth (Albimi; obtainable from Micro-Bio Laboratories Ltd., London) is prepared. For a semi-solid medium Agar No. 3 (Oxoid) is added to give a final concentration of 0.1% (w/v). Methylene blue is sometimes added to the semi-solid medium to a final concentration of 0.002% (w/v) as an oxidation—reduction indicator. Various other additions may be made to ABB for use in differential tests (Park, Munro, Melrose and Stewart, 1962). Vibrios may grow in peptone water ($\%$, w/v, in demineralized H_2O; Evans peptone, 1.0; NaCl, 0.5; final pH, 7.0) if it is boiled immediately before inoculation and incubated in 30% (v/v) CO_2 in air.

Gaseous conditions

The gaseous requirements of *C. fetus* and related organisms have been studied by several workers, most of whom have incubated at atmospheric pressure and have concluded that the oxygen tension of the atmosphere should be lower and the CO_2 tension higher than it is in air. For example, Kiggins and Plastridge (1956) reported that the optimal oxygen concentration for *C. fetus* growing on blood agar was 5% (v/v) and that this concentration could be achieved by diluting air with H_2, N_2 or CO_2. Their data showed that there was little difference in amount of growth of vibrios in 5% (v/v) O_2 compared with in 15% (v/v) O_2 when CO_2 was the diluent used. The optimum CO_2 concentration was in the range 5–30% (v/v). They found that *C. fetus* did not grow in air or anaerobically. Several workers have used other gaseous conditions. Dennis and Jones (1959) bubbled air diluted with N_2 to give a final concentration of 6% (v/v) O_2 through a liquid medium which contained peptone, yeast extract and cysteine and which was used for the bulk cultivation of *C. fetus*. Firehammer (1965) grew his isolates in air at reduced pressure (270 Torr). Garvie (1967) obtained evidence which suggested that the presence of blood in a medium altered the gaseous requirements of vibrios.

We have found that satisfactory growth of microaerophilic vibrios is obtained by incubating TBA in an atmosphere of 30% (v/v) CO_2 in air. This atmosphere is achieved by reducing the pressure within anaerobic jars, or other suitable containers, by one third and then using CO_2 to replace the air thus removed. A Kipps apparatus can be used to produce the gas, but CO_2 prepared in this way should be passed through a saturated solution of $NaHCO_3$ to remove any HCl fumes.

Semi-solid media are incubated in air with the caps of the bottles tightened.

Temperature

Microaerophilic vibrios grow well at 37° and this is the incubation temperature generally used. King (1962) showed that the temperature range for growth varies between strains of vibrios and she used this characteristic for differentiation.

Maintenance and Preservation of Cultures

Cultures can usually be maintained in Brewer's medium with the agar concentration increased to 0·4% (w/v) but it is advisable also to keep them in another medium such as Robertson's cooked meat. After incubation for 3 days at 37° the culture bottles are stored unopened at room temperature until required. Care should be taken not to disturb the growth in Brewer's medium as this frequently reduces viability. Subcultures should be made at least monthly. Microaerophilic vibrios have been preserved by freeze-drying, but survival is poor unless special methods are used for recovery (Garvie, 1967).

Isolation

Several selective media for microaerophilic vibrios have been described (e.g. Terpstra, 1954; Florent, 1956; Kuzdas and Morse, 1956; Plastridge, Koths and Williams, 1961). However, following Morgan (1957) and Morgan, Melrose and Stewart (1958) we use TBA and rely on recognizing colonies or spreading growth of vibrios. Differential filtration of samples that are likely to be heavily contaminated is sometimes used to remove organisms other than vibrios (see below). Actidione (Sigma, London) is incorporated in the medium to a final concentration of 0·01 mg/ml when contamination by moulds is a problem. After inoculation plates are incubated in an atmosphere of air (70%, v/v) and CO_2 (30%, v/v) and are examined after 3 and 7 days. Inclusion of absorbent paper or dried silica gel in the containers used for incubation serves to reduce the amount of moisture on the plates.

Isolation from the genital tract of cattle

Examination of the genital tract of cows for microaerophilic vibrios involves the removal of a sample of vaginal mucus to a plate of TBA, the sample being spread over the plate with a glass spreader. The same spreader is rubbed over three other plates so that a set of plates with a range of inoculum sizes is produced.

Bulls, especially those used for artificial insemination, should be tested periodically to ensure that they do not carry *C. fetus*, since this organism causes an infectious infertility. Although *C. bubulus* can be readily isolated from semen and preputial washings, isolation of *C. fetus* is difficult. Thus test mating is generally used for the diagnosis of vibriosis in bulls (see Lawson, 1959). This involves insemination of a heifer that has been shown to be free from *C. fetus* infection. Vaginal mucus from the heifer is then examined twice weekly for *C. fetus*. Regular recovery of the organism over one month is taken to show that the bull is a carrier of a strain of *C. fetus* capable of multiplication in the female genital tract. For this reason the technique is suitable for detecting strains likely to be a cause of infertility. The fluorescent antibody technique has been applied to the detection of *C. fetus* in bulls and would seem to offer an attractive alternative method of diagnosis (Philpott, 1968a, b).

Isolation from aborted foetuses

A sample of fluid from the foetal stomach is transferred to a plate of TBA which is then incubated. A pure culture is usually obtained if *C. fetus* was the cause of abortion and if the foetus has not been grossly contaminated. When contamination is suspected, cultivation from the brain should be attempted.

Isolation from man

Blood, or material from lesions may be streaked on TBA plates. Blood may also be examined by blood culture followed by streaking on TBA. Organisms like *C. bubulus* have been isolated from the mouth by using a membrane filtration technique (Loesche, *et al*, 1965).

Isolation from the gut of pigs

At *post mortem* examination portions of gut *c.* 40 mm long are removed and placed in Petri dishes. Each portion is then cut along its length, the majority of the gut contents is removed and the portion of gut then put in

Q

¼ strength Ringer's solution (10 ml) in a 1 oz McCartney bottle. The bottle is shaken to remove any remaining contents from the gut portion, and the portion is then put in a sterile Petri dish, lumen surface uppermost. The lumen surface is seared with a hot scalpel and the seared material is removed and discarded. A sample is then removed from the sub-mucosal region, by scraping with a small scalpel, and transferred to a plate of TBA. A wire loop or glass spreader is used to distribute the inoculum over the plate and is then used, without re-sterilizing, to inoculate three further plates so that a set of four plates with a range of inoculum sizes is obtained.

Isolation from gut contents and faeces

Samples (1 g) of gut contents or faeces are suspended in ¼ strength Ringer's solution (10 ml) in a 1 oz Universal bottle. After allowing to stand for 30 min so that the coarse material settles, a sample is removed from the upper layer of the suspension and is passed through a membrane filter (Millipore MF type, DA; mean pore size 0·65 μm) used in conjunction with a pre-filter (Millipore AP20). A convenient method is to withdraw the sample from the bottle into a 5 ml syringe, remove the needle, fit a micro syringe filter holder (Millipore, XX30,02500) plus prefilter and filter, and discharge 2 drops (C. 0·05 ml) of the sample, through the filter, on to a TBA plate. The drops are then spread over the plate with a glass spreader. The membrane filtration method has been used by several workers for isolating microaerophilic vibrios (Plumer, Duvall and Shepler, 1962; Loesche et al., 1965; Seamon, 1968).

Isolation of vibrios from the rectum of healthy pigs can be achieved by swabbing the area and then, within an hour, smearing the swab over the surface of a TBA plate. A wire loop is then streaked across this plate and used to inoculate three more plates.

Identification

On isolation plates, especially when they are moist, vibrios are often seen as thin translucent bluish-grey films. Three colony types commonly found on TBA after incubation for 3 days at 37° are:

(1) round, convex, shiny, entire edge; light grey; diam 2 mm (Fig. 1a).

(2) irregular, tending to spread, raised, slightly granular, entire or undulate edge; light grey; diam 5 mm or more (Fig. 1b).

(3) round, raised, shiny (granular centre in large colonies), entire edge; light grey at edge of colony becoming cream towards the centre, larger colonies have slightly green coloration; diam 3–5 mm (Fig. 1c).

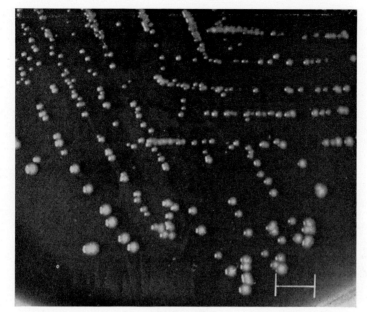

Fig. 1. (a)

Fig. 1. (b)

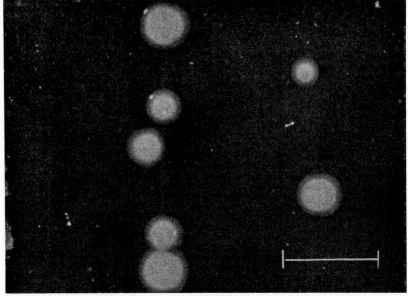

FIG. 1. (c)

FIG. 1. Three colony types of microaerophilic vibrios common on TBA after incubation for 3 days. (bar = 10 mm).

When vibrios are examined by Gram's method they appear as Gram-negative slender curved or S-shaped rods and, especially in some cultures, as spirals and long wavy forms. The organisms may not stain strongly so it is often useful to counterstain for longer than the normal period and to rinse and dry the slide as quickly as possible. In Gram-stained preparations vibrios appear to be $c.$ 0·4 μm wide. The curved rods are 1–3 μm long and the spirals up to 20 μm long. Old cultures contain many spherical forms and when examined at this stage may not be recognized as cultures of vibrios. When in doubt it is advisable to examine a subculture after incubation for 2 or 3 days. Vibrios are motile by a flagellum at one or both poles.

Campylobacter spp do not attack carbohydrates, hydrolyse gelatin, starch, urea or lipid, haemolyse blood or produce indole. Most workers agree that nitrate is reduced to nitrite and that at least two species can be differentiated; *C. fetus*, which is catalase positive and which produces little or no H_2S detectable by lead acetate paper when grown in a medium without added cysteine, and *C. bubulus*, which is catalase negative and produces abundant H_2S. As stated earlier, further subdivisions within *Campylobacter* have been suggested but the detailed systematics of the genus does not yet seem to be clear. For details of various differential tests applied to *Campylobacter*

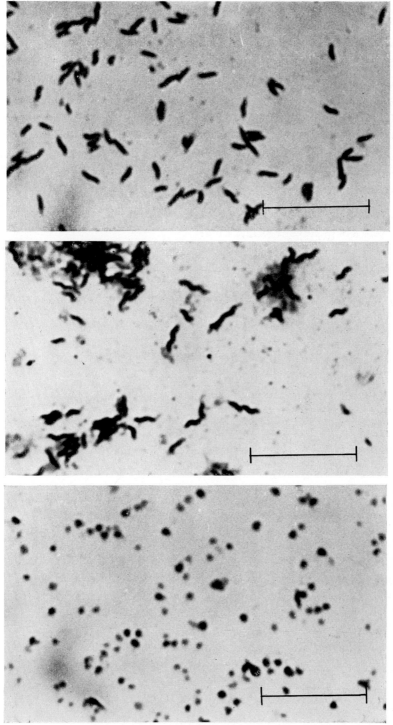

FIG. 2. Morphological forms of microaerophilic vibrios from TBA as seen on Gram-stained smear (bar = 10μm).

consult Florent (1959), King (1962), Park *et al.* (1962) and Loesche *et al.* (1965).

Acknowledgements

We wish to thank for their help in this work: Dr N. S. Barron, Agriculture Department, and Mr P. Cockburn, Upperwood Farm, Reading University; Mrs A. Gush and Mr R. M. Loosemore, Veterinary Investigation Centre, Coley Park; Mr A. D. Osborne, Department of Veterinary Medicine, Bristol University. Our thanks are also due to Mr M. Crowder for his conscientious technical assistance.

We gratefully acknowledge a grant from the Meat and Livestock Commission in support of this work.

References

BASDEN, E. H., TOURTELLOTTE, M. E., PLASTRIDGE, W. N. & TUCKER, J. S. (1968). Genetic relationship among bacteria classified as vibrios. *J. Bact.,* **95,** 439.

DENNIS, S. M. & JONES, R. F. (1959). A method for bulk growth of *Vibrio fetus. Aust. vet. J.,* **35,** 457.

DOYLE, L. P. (1948). Etiology of swine dysentery. *Am. J. vet. Res.,* **9,** 50.

FIREHAMMER, B. D. (1965). The isolation of vibrios from ovine feces. *Cornell Vet.,* **55,** 482.

FLORENT, A. (1953). Isolement d'un vibrion saprophyte du sperme du taureau et du vagin de la vache (*Vibrio bubulus*). *C. R. Soc. Biol., Paris,* **147,** 2006.

FLORENT, A. (1956). Methode d'isolement de *Vibrio foetus* a partir d'echantillons polymicrobiens, specialement du liquide preputial. Milieu selectif "coeur-sang-gelose au vert brilliant" en microaerobiose. *Soc. Belge. de Biol., Compt. Rend.,* **150,** 1059.

FLORENT, A. (1959). Les deux vibrioses génitales: la vibriose due a *V. fetus venerealis* et la vibriose d'origine intestinale due a *V. fetus intestinalis. Med. Veeartsenijschool, Ghent,* **3,** part iii.

FLORENT, A. (1963). A propos des vibrions responsables de la vibriose génitale des bovins et des ovins. *Bull. Off. int. Epiz.,* **60,** 1063.

GARVIE, E. I. (1967). Preservation of *Vibrio fetus* by freeze drying. *J. appl. Bact.,* **30,** 255.

JONES, F. S., ORCUTT, M. & LITTLE, R. B. (1931). Vibrios (*Vibrio jejuni* n. sp.) associated with intestinal disorders of cows and calves. *J. exp. Med.* **53,** 853.

KIGGINS, E. M. & PLASTRIDGE, W. N. (1956). Effect of gaseous environment on growth and catalase content of *Vibrio fetus* cultures of bovine origin. *J. Bact.,* **72,** 397.

KING, E. O. (1962). The laboratory recognition of *Vibrio fetus* and a closely related *Vibrio* isolated from cases of human vibriosis. *Ann. N. Y. Acad. Sci.,* **98,** 700.

KUZDAS, C. D. & MORSE, E. V. (1956). A selective medium for the isolation of *Vibrio fetus* and related vibrios. *J. Bact.,* **71,** 251.

LAWSON, J. R. (1959). Vibriosis. In Stableforth, A. W. & Galloway, I. A. *Infectious Diseases of Animals. Diseases due to Bacteria.* London: Butterworths, p. 745.

LOESCHE, W. J., GIBBONS, R. J. & SOCRANSKY, S. S. (1965). Biochemical characteristics of *Vibrio sputorum* and relationship to *Vibrio bubulus* and *Vibrio fetus. J. Bact.,* **89,** 1109.

MORGAN, W. J. B. (1957). A note on the isolation and identification of *Vibrio fetus* from cattle. *Vet. Rec.,* **69,** 32.

MORGAN, W. J. B., MELROSE, D. R. & STEWART, D. L. (1958). The diagnosis of *Vibrio fetus* infection in bulls. *Vet. Rec.,* **70,** 93.

PARK, R. W. A. (1961). A note on the systematic position of *Vibrio fetus. J. appl. Bact.,* **24,** 23.

PARK, R. W. A., MUNRO, I. B., MELROSE, D. R. & STEWART, D. L. (1962). Observations on the ability of two biochemical types of *Vibrio fetus* to proliferate in the genital tract of cattle and their importance with respect to infertility. *Br. vet. J.,* **118,** 411.

PHILPOTT, M. (1968a). Diagnosis of *Vibrio fetus* infection in the bull. I. A modification of Mellick's fluorescent antibody test. *Vet. Rec.,* **82,** 424.

PHILPOTT, M. (1968b). Diagnosis of *Vibrio fetus* infection in the bull. II. An epidemiological survey using a fluorescent antibody test and comparing this with a cultural method. *Vet. Rec.,* **82,** 458.

PLASTRIDGE, W. N., KOTHS, M. E. & WILLIAMS, L. F. (1961). Antibiotic mediums for the isolation of vibrios from bull semen. *Am. J. vet. Res.,* **22,** 867.

PLUMER, G. J., DUVALL, W. C. & SHEPLER, V. M. (1962). A preliminary report on a new technique for isolation of *Vibrio fetus* from carrier bulls. *Cornell Vet.,* **52,** 110.

PREVOT, A. R. (1940). Classification des *Vibrions* anaérobies. *Annls Inst. Pasteur, Paris,* **64,** 117.

SEAMON, P. J. (1968). A membrane filtration technique for the isolation of *Vibrio* species from contaminated material. *J. med. Lab. Technol.,* **25,** 25.

SEBALD, M. & VÉRON, M. (1963). Teneur en bases de l'ADN et classification des vibrions. *Annls Inst. Pasteur, Paris,* **105,** 897.

SMIBERT, R. M. (1963). Nutrition of *Vibrio fetus. J. Bact.,* **85,** 394.

SMITH, T. & TAYLOR, M. S. (1919). Some morphological and biological characters of the spirilla (*Vibrio fetus* n. sp.) associated with disease of the fetal membranes in cattle. *J. exp. Med.,* **30,** 299.

TERPSTRA, J. I. (1954). Vibriose bovine. *Bull. Off. int. Épiz.,* **42,** 641.

The Detection of *Zymomonas anaerobia*

M. J. S. DADDS

Process Research Department, Allied Breweries Ltd.
Burton-on-Trent, Staffordshire, England

Zymomonas anaerobia is a Gram-negative, unicellular, straight non-sporing rod, up to $2\,\mu$ in width and 3–$4\,\mu$ in length (up to $10\,\mu$ in old cultures). It is motile for upwards of 6 days following inoculation and forms rosette-like clusters during the motile phase. The non-motile strain *Z. anaerobia* var *immobilis* does not form these characteristic clusters, but is similar in all other characteristics to *Z. anaerobia* (Shimwell, 1937, 1950). The organism has been described as anaerobic to microaerophilic. Only glucose and fructose are used as energy sources for growth, Millis (1956), Dawes and Large (1968) and Dadds (unpublished observations). These carbohydrates are fermented via the Entner-Doudoroff pathway yielding ethanol, lactic acid, carbon dioxide and traces of acetaldehyde (MacGee and Doudoroff, 1954). There is some disagreement about the vitamin requirements, for whilst it has been reported that biotin and lipoic acid stimulate growth (Dawes and Large, 1968) I have found that *Z. anaerobia* has an absolute requirement for pantothenate. *Zymomonas mobilis* also has a requirement for this vitamin (Belaïch and Senez, 1965). In addition *Z. anaerobia* cannot grow appreciably in the absence of amino acids (Dawes and Large, 1968) and noticeably its growth in some beers is negligible without the addition of yeast extract (Dadds, unpublished observations).

Beer infected with this organism is characterized by turbidity, "fruity flavours" and a distinct sulphidic aroma accompanied by relatively high levels (50–90 ppb) of free hydrogen sulphide (M. J. S. Dadds, unpublished observations). The fruity flavours are probably due to the production of acetaldehyde which, incidentally, combines with sulphur dioxide, the only preservative allowed in beer (Burroughs and Sparks, 1964).

Z. anaerobia var *pomaceae* has been identified as one of the causal organisms of "cider sickness" (Millis, 1956). Characteristically the symptom of an infected cider is a change from the normal sluggish fermentation to one of "explosive vigour" accompanied by an aroma of acetaldehyde or raspberries (Pollard, 1959) and a marked turbidity. This turbidity is due to large numbers of bacterial cells and a precipitate of a tannin aldehyde complex (Barker, 1948).

The Differential Medium

Per litre distilled water:

Malt Extract (Oxoid)	3·0 g
Yeast Extract (Oxoid)	3·0 g
Glucose	20·0 g
Peptone	5·0 g
"Acti-dione"	20 mg (20 ppm)

The pH is adjusted to 4·0 with dilute hydrochloric or sulphuric acid. The medium is dispensed into screw-capped bottles (20 ml/bottle) and then sterilized by autoclaving at 121° for 15 min. Filter sterilized ethanol (0·6 ml, 95% v/v) is added aseptically to each bottle giving a final concentration of $\approx 3\%$ (v/v).

The test

Samples of infected beer (1 ml) are added to the differential medium (20 ml) in a screw-capped bottle (capacity 25 ml) containing an inverted Durham tube (40 × 7 mm). The presence of *Zymomonas* is indicated by a Durham tube full of gas (CO_2) and by a marked effervescence. Further identification is aided by Gram-staining and microscopic examination.

A positive result can be obtained after 2–6 days at 25° depending on the size of the inoculum. One to two "infecting units"/ml can be detected provided that the proper precautions are taken. The term "infecting unit" must be used because 40–50% of the cells occur as pairs.

To detect low levels of *Zymomonas* in beer an enrichment stage is required. Priming solution is added to a pint or a half pint (234 ml) of beer giving the following final concentrations: Glucose 2% w/v, yeast extract ("Oxoid") 0·1% w/v and "Acti-dione" (Cycloheximide, The Upjohn Company, Kalamazoo, Michigan) 20 ppm w/v.

The supplemented beer is incubated at 25–30° in a bottle with a loose-fitting cap. If *Zymomonas* is present the beer will eventually become turbid accompanied by the characteristic sulphidic aroma. This can then be tested for *Zymomonas* using the differential medium, and a positive result is usually obtained after 1–2 days.

Specificity of the test

Of the bacteria normally found in infected brewery materials only four groups might produce false positive (visible gas production) results.

Normally, the anaerobic conditions in the medium prevent the growth of *Acetobacter*, whilst the addition of ethanol (3% v/v) controls the growth of *Klebsiella aerogenes* and *Enterobacter aerogenes*. Most strains of the *Lactobacillus brevis—pastorianus* group found in spoiled beer produce negligible quantities of gas. However a few isolates do produce relatively large quantities of gas and these can be identified by Gram-staining and by their microscopic appearance. It has not been found possible to control the growth of these lactics without reducing the sensitivity of the test.

"Acti-dione" (20 ppm w/v) inhibits gas production by *Saccharomyces cerevisiae* provided that the inoculum does not contain more than 2×10^6 cells. "Wild yeasts" (non-brewing strains of *Saccharomyces* viz: *Sacch. pastorianus*, *Sacch. cerevisiae* var *ellipsoideus*, *Sacch. marxianus* and *Sacch. fragilis*) are not inhibited above a concentration of $0 \cdot 2 – 0 \cdot 4 \times 10^6$ cells.

"Mycostatin" (Nystatin, E. R. Squibb and Sons, Liverpool and New York) (10 units/ml) and "Fungizone" (Amphotericin B, E. R. Squibb and Sons, Liverpool and New York) (5 μg/ml) will inhibit gas production from an inoculum of up to $1 \cdot 5 – 2 \times 10^6$ cells of most yeasts found in the brewing environment.

Whilst these polyene antibiotics are more effective than "Acti-dione" they are heat labile (95% loss in activity during autoclaving at 121° for 15 min at pH 4·0), and are more expensive. These factors preclude their use in most quality control laboratories.

Neither "Mycostatin", "Fungizone" nor "Acti-dione" influence the sensitivity of the test at the levels used; however the concentration of ethanol must be maintained at 3% v/v $\pm$ 0·2% for at 4% the sensitivity is noticeably reduced.

The diluent

The choice of diluent used for swabbing and dilution plate counts is important. The addition of glucose 1% strength Ringer's solution maintains the viability at a higher level than with Ringer's solution alone.

Acknowledgements

The author wishes to thank the directors of Allied Breweries Ltd. for permission to publish this paper and Professor A. H. Rose for his help and encouragement.

References

BARKER, B. T. P. (1948). Some recent studies on the nature and incidence of cider sickness. *Rep. R. agric. hort. Res. Sta. Bristol,* 174.

BELAICH, J. P. & SENEZ, J. C. (1965). Influence of aeration and of pantothenate on growth yields of *Zymomonas mobilis*. *J. Bact.*, **89**, 1195.

BURROUGHS, L. F. & SPARKS, A. H. (1964). The determination of free sulphur dioxide content of ciders. *Analyst*, **89**, 55.

DAWES, E. A. & LARGE, P. J. (1968). The endogenous metabolism of anaerobic bacteria. *U.S. Govt., Res. Develop. Rep.*, **68**, (23), 50, *Report No. AD* 676181.

MACGEE, J. & DOUDOROFF, M. (1954). New phosphorylated intermediate in glucose oxidation. *J. biol. Chem.*, **210**, 617.

MILLIS, NANCY, F. (1956). A Study of the cider sickness bacillus—a new variety of *Zymomonas anaerobia*. *J. gen. Microbiol.*, **15**, 521.

POLLARD, A. (1959). Le framboisé du cidre et la "cider sickness". *Ind Aliment. Agricol.*, 537.

SHIMWELL, J. L. (1937). Study of a new type of beer disease bacterium *Achromobacter anaerobium* spec. nov. producing alcoholic fermentation of glucose. *J. inst. Brew.*, **56**, 179.

SHIMWELL, J. L. (1950). *Saccharomonas:* a proposed new genus for bacteria producing a quantitative alcoholic fermentation of glucose. *J. inst. Brew.*, **56**, 179.

The Isolation and Enumeration of Sulphate-Reducing Bacteria

Eileen S. Pankhurst

The Gas Council, London Research Station, London, England

The media and techniques described here apply to the **dissimilatory** sulphate-reducing bacteria, first isolated and described over seventy years ago by Zelinsky (1893), Beijerinck (1895) and ven Delden (1903). Since then, these ubiquitous organisms have been subjects of interest and research, not least because of their unusual mode of life: they obtain energy by reduction of the sulphate ion which acts as the terminal electron acceptor in an anaerobic process. Although certain other organisms are capable of reducing smaller quantities of sulphate, they do so only in an **assimilatory** manner, to obtain sulphur for syntheses of cell material.

Dissimilatory sulphate-reducing bacteria are now classified in two genera, described by Campbell and Postgate (1965) and by Postgate and Campbell (1966).

Genus *DESULFOVIBRIO*	Genus *DESULFOTOMACULUM*
D. desulfuricans	*D. nigrificans**
D. vulgaris	*D. orientis*
D. salexigens	*D. ruminis*
D. africanus	
D. gigas	

Desulfovibrio spp are anaerobic, do not sporulate, usually have a polar flagellum, contain cytochrome C_3 and also generally contain desulfoviridin, a green pigment of unknown function; the percentages of guanine and cytosine in their DNA are usually higher than for *Desulfotomaculum* spp.

Desulfotomaculum spp are either mesophilic or thermophilic, sporulate, have peritrichous flagella, contain cytochrome b, but not desulfoviridin; the percentages of guanine and cytosine in their DNA are lower than for *Desulfovibrio* spp.

This chapter deals mainly with the isolation and enumeration of mesophilic sulphate-reducing bacteria, as the author has little experience with

* Formerly known as *Clostridium nigrificans*.

thermophilic strains. The same techniques can be applied to thermophils, but, for reasons which are by no means obvious, not as effectively. In fact, Bufton (1959) concluded that there was no satisfactory way of counting *Desulfotomaculum nigrificans*. Isolation of the rumen sulphate-reducer (*D. ruminis*) requires special techniques, but once isolated it can be maintained in the same media as other sulphate-reducing organisms; for techniques specific to rumen bacteria, the reader is referred to other chapters in this book, and to the report of the isolation of *D. ruminis* (Coleman, 1960).

Sulphate-reducing bacteria have long had the reputation of being difficult to grow and enumerate, and this reputation is enhanced by the fact that the procedures involved can be tedious and time-consuming. However, both pure and mixed cultures are relatively easy to grow under anaerobic, strongly reducing conditions. Moreover, if such conditions are maintained, these organisms are also remarkably long-lived.

Media

Six of the media most commonly used for isolating and growing sulphate-reducing bacteria are described in Table 1; most are used as liquids but can be solidified by agar if required. Under anaerobic conditions, sulphate-reducing bacteria readily grow in liquid media and form colonies within solid and semi-solid media. Yet their growth on the surface of agar is usually described as sparse and erratic. However, *D. vulgaris* can grow well (Pankhurst, 1966b, 1967a) on Baars' medium containing yeast extract, sodium thioglycollate and Ionagar. Good growth of *Desulfovibrio* spp has also been obtained on yeast extract agar and trypticase soy agar, accompanied by production of a growth-stimulating gas believed to be a low oxide of sulphur (Iverson, 1966, 1967).

Many present-day media are essentially the same as the early ones, but with two most important modifications: addition of small quantities of yeast extract or a similar substance, and the use of reducing agents to control the initial oxidation-reduction potential (E_h). As the function of yeast extract in the medium and E_h control are important, they will be discussed briefly. Yeast extract and sodium thioglycollate in Baars' medium can decrease by a factor of 10^4 the number of cells required to initiate growth (Pankhurst, 1967a). The stimulatory effect of yeast extract, described by a number of workers including Bunker (1939), Butlin, Adams and Thomas (1949) and Miller (1949), is partly due to constituent amino acids. Three of these, serine, ornithine and isoleucine, can be replaced by the chelating agent EDTA; they probably increase the solubility of ferrous sulphide and thus the availability of inorganic iron (Postgate, 1951, 1953, 1965).

But yeast extract also plays a part in the nutrition of sulphate-reducing organisms. Mechalas and Rittenberg (1960) found that cells grown in the presence of yeast extract and radio-active sodium bicarbonate obtained most of their carbon from the former. Thus the veracity of previous reports by Starkey and Wight (1945) and by Butlin, Adams and Thomas (1949) that sulphate-reducing bacteria were autotrophic (i.e. capable of growth without an exogenous source of organic carbon) was questioned, especially as Mechalas and Rittenberg were unable to grow these organisms without yeast extract. In the earlier work, organic impurities in the inorganic media were probably sufficient to support growth (Postgate, 1960). Yet during heterotrophic growth, some *Desulfovibrio* strains assimilate 3–4 times as much carbon dioxide as other heterotrophic bacteria (Postgate, 1960; Sorokin, 1960, 1961). One strain (Sorokin, 1966) can even obtain 50% of its cell carbon from carbon dioxide, if the medium contains trace amounts of acetate. Thus although sulphate-reducing bacteria can assimilate significant amounts of carbon dioxide, it is now accepted that they do not grow as strict autotrophs, but need small quantities of organic carbon.

The significance of low E_h values in highly corrosive soils (Starkey and Wight, 1945), and of the beneficial effects of adding reducing agents to media (Baars, 1930; Butlin *et al.*, 1949), was not at first appreciated. Then Grossman and Postgate (1953a, b) showed that L-cysteine hydrochloride, sodium sulphide and thioglycollic acid (sometimes referred to as mercapto-acetic acid or thiolacetic acid) did not function in a nutritive capacity, but established a low E_h in the medium. Although some media already contain reducing compounds, and organisms and sulphide in large inocula tend to create their own reducing conditions, most media require the addition of one or more specific reducing agents to establish E_h values low enough for growth of sulphate-reducing bacteria.

Normally, the E_h in an uninoculated medium should be below 0, and for small inocula it should be about -200 mv Postgate (1959). For freshly-prepared Baars' medium containing yeast-extract and thioglycollate, the E_h is usually *c.* -150 mv, falling to -200 to -250 mv after growth of *D. vulgaris* (Pankhurst, 1967a). Sodium sulphide in the chemically-defined medium described in Table 1 gives an initial potential of -300 mv (MacPherson and Miller, 1963), but its use is limited to media in which precipitation of ferrous sulphide is not used as a criterion of growth. Thioglycollic acid, inhibitory at high concentrations, has been used successfully by Allred, Mills and Fisher (1954), by Booth and Wormwell (1961), and by Adams and Postgate (1959), who found it better than cysteine for counts of *D. orientis.*

A disadvantage of using cysteine in some media is that precipitation of black ferrous sulphide, which usually means that sulphide has been formed

TABLE 1. Some media; for sulphate-reducing bacteria for halophilic strains, or *Baars' medium*. Baars, 1930; modified by the addition of yeast extract (Butlin, Adams and Thomas, 1949), by the addition of thioglycollate (Grossman and Postgate, 1953*a*, and *b*), and by the use of Ionagar (Pankhurst, 1967*a*).

Very useful medium for detection, cultivation and enumeration. Used in liquid or agar form.

K_2HPO_4	0·5g
....	
NH_4Cl	1·0g
$CaSO_4$	1·0g
$MgSO_4.7H_2O$	2·0g
Sodium lactate (70% w/w soln)	5·0g
$FeSO_4(NH_4)_2.SO_4.6H_2O$	0·5g
Yeast extract (Oxoid powder)	1·0g
....	
Sodium thioglycollate	0·1g
....	
'Ionagar" No 2 (Oxoid)	10·0g
Tap water	1 litre
pH	7·4 ± 0·2

Prepare in 930 ml water, without ferrous salt, yeast extract and thioglycollate; adjust pH to 8·1. Autoclave for 20 min at 115°, cool rapidly to prevent re-solution of oxygen. Then add: 50 ml supernatant liquid from 1% (w/v) aq. solution of the ferrous salt (steam-sterilized for 1 h on 3 successive days); 10 ml of 10% (w/v) aq. solution of yeast extract (autoclave and membrane filter); 10 ml of 1% (w/v) aq. solution of sodium thioglycollate (filter sterilized). Final pH usually 7·2–7·6 without further adjustment. Disperse sediment with medium into culture vessels.

Medium C (Postgate, 1966, 1969a; similar to media used by Starkey (1938) and Butlin, Adams and Thomas (1949)).

Particularly useful for growing cells in continuous culture, for biochemical work and for manometric studies. It contains no sediment and ferrous sulphide is not precipitated during growth. Normally used in liquid form.

....	
KH_2PO_4	0·5g
NH_4Cl	1·0g
Na_2SO_4	4·5g
$CaCl_2.6H_2O$	0·06g
$MgSO_4.7H_2O$	0·06g
Sodium lactate	6·0g
$FeSO_4.7H_2O$	0·004g
Yeast extract	1·0g
Sodium citrate. $2H_2O$	5·0g
....	
....	
Distilled water	1 litre
pH	7·5 ± 0·2

Sterilize by autoclaving 20 min. at 115°. Maybe cloudy after autoclaving but should clear on cooling; pH tends to become alkaline when growth occurs – remains steadier if ammonium sulphate (7g) is used instead of sodium sulphate. Na_2S can be added aseptically to encourage the growth of small inocula.

Medium D (Postgate, 1966, 1969a).

For testing ability to grow with choline or sodium pyruvate, in the absence of sulphate. Used in liquid form.

....	
KH_2PO_4	0·5g
NH_4Cl	1·0g
Sodium pyruvate (or choline chloride)	1·0g
$CaCl_2.6H_2O$	0·1g
$MgCl_2.6H_2O$	1·6g
....	
$FeSO_4.7H_2O$	0·004g
Yeast extract	1·0g
....	
....	
....	
Distilled water	1 litre
pH	7·5 ± 0·2

Sterilized by membrane filtration.

samples from salt water environments, media should contain 2·5 % (w/v) NaCl.

Medium E. (Postgate, 1966, 1969*a*).

For enumeration. Used in agar form.

KH$_2$PO$_4$	0·5g
NH$_4$Cl	1·0g
Na$_2$SO$_4$	1·0g
CaCl$_2$.6H$_2$O	1·0g
MgSO$_4$.7H$_2$O	2·0g
Sodium lactate	3·5g
FeSO$_4$.7H$_2$O	0·5g
Yeast extract	1·0g
Thioglycollic acid	1·0g
Ascorbic acid	1·0g
Agar	15·0g
Tap water	1 litre
pH	7·6

Ingredients dissolved by boiling; adjust pH to 7·6 with NaOH. Sterilize by autoclaving 20 min. at 115°. The thioglycollic acid should be sterilized separately by membrane filtration.

A.P.1. medium (American Petroleum Institute, 1965).

For detection, cultivation and enumeration. Used in liquid or agar form.

K$_2$HPO$_4$ (anhydrous)	0·01g
NaCl	10·0g
MgSO$_4$.7H$_2$O	0·2g
Sodium lactate	4·0ml
FeSO$_4$(NH$_4$)$_2$SO$_4$.6H$_2$O	0·2g
Yeast extract	1·0g
Ascorbic acid	0·1g
Agar	15·0g
Distilled water	1 litre
pH	7·3

Dissolve ingredients with gentle heating; adjust pH to 7·3; sterilize by autoclaving for 10 min at 121°. For the liquid version, add the iron salt after the rest of the medium has been sterilized.

Chemically-defined medium (MacPherson and Miller, (1963); formula originally given in molar concentrations).

For nutritional studies.

KH$_2$PO$_4$	0·34g
NH$_4$Cl	0·53g
Na$_2$SO$_4$	7·10g
CaCl$_2$	0·06g
MgSO$_4$.7H$_2$O	0·06g
Lactic acid	9·01g
FeSO$_4$.7H$_2$O	0·007g
Trace elements B,Co,Cu,Mn,Mo,Zn	0·0005g each
Na$_2$S	0·08g
Distilled water	1 litre
pH	7·3 ± 0·1

Make up medium without ferrous salt and sulphide; adjust pH to 6·5 with NaOH. Sterilize by autoclaving 20 min 'at 115°. When cool, add the ferrous salt (membrane filtered solution in 2·5 mM. H$_2$SO$_4$) and adjust the pH to 7·2–7·4. Add the Na$_2$S before inoculation so that the redox potential is *c*.-300mv.

R

from sulphate by sulphate-reducing organisms, can occur because sulphide is liberated from cysteine by organisms that decompose this amino acid. To eliminate this possibility, subcultures are made into media without cysteine, or samples are examined microscopically for typical sulphate-reducing bacteria (Grossman and Postgate, 1953b). Thioglycollic acid also contains an SH group but organisms that form hydrogen sulphide from it are uncommon. Two other reducing agents sometimes used are ascorbic acid and glutathione. Iron nails have also been employed to great effect (Abd-el-Malek and Rizk, 1958, 1960). In the presence of metallic iron, the potential of uninoculated media containing sodium lactate and inorganic salts can be as low as -280 to -400 mv (Starkey and Wight, 1945), and usually alters little, even during growth.

Preparation

Notes on the preparation and uses of media are given in Table 1; Analar chemicals should be used whenever possible. Most media should be prepared just before use, and pH values should be checked after preparation. If there is any doubt, E_h values should be checked also. A suitable combined platinum/calomel electrode (W. G. Pye & Co. Ltd., Cambridge, England) is immersed in the medium and left to equilibrate for a minute or two; the medium is mixed well, but, of course, not aerated. The potential of the sample relative to the calomel half cell is read in mv from a pH meter; conversion to E_h (the potential relative to the hydrogen electrode) is made by adding the appropriate factor. After use, the platinum part of the electrode should be polished carefully with a soft cloth and cleaned, if necessary, with dilute HCl. For more information on E_h measurement, the reader should consult Hewitt (1950).

We occasionally have to sterilize yeast extract by membrane filtration, as sporing organisms in some batches survive autoclaving and subsequently cause contamination. Postgate (1969a) commented on occasional difficulties when thioglycollate is added to media before they are autoclaved. Although we have not experienced these particular difficulties, we find that commercial batches of sodium thioglycollate vary. Sometimes they are already decomposed and smell strongly of H_2S; they should not be used in this condition.

Although most media deteriorate quickly after preparation and cannot therefore be stored, time can be saved by weighing ingredients beforehand, and keeping them separate and labelled, in the dark; they can then be dissolved and mixed when the medium is required. Sodium thioglycollate, cysteine and other constituents that oxidize or decompose readily must not be stored in this way however. Concentrated batches of sterile yeast extract can also be prepared in advance. A little foresight does much to relieve the tedium of growing and counting sulphate-reducing bacteria.

Conditions of incubation

Anaerobic cultures must be incubated so that access of oxygen is prevented and reducing conditions for growth provided. The inhibitory effect of oxygen on sulphate-reducing bacteria was established long ago; reports to the contrary, and survival of organisms in oxidizing conditions (Rogers, 1940; Zobell, 1958) can probably be explained by the existence of anaerobic micro-environments, in clumps of mucoid material or ferrous sulphide for example. When air is bubbled slowly through cultures of D. vulgaris no growth occurs, and after several days, there are fewer viable cells; more vigorous aeration on a shaking machine kills cells within two to three hours (Pankhurst, 1967a). Some sulphate-reducing organisms contain the enzyme hydrogenase and therefore can reduce sulphate with molecular hydrogen. Hydrogenase is oxidized if oxygen is passed through cultures of bacteria containing it, or through cell-free preparations; it can be reactivated if cells are incubated in hydrogen, or in the case of enzyme preparations, if oxygen is removed physically, chemically or enzymatically (Hoberman and Rittenberg, 1943; Fisher, Krasna and Rittenberg, 1954). In some cases, sulphate-reducing bacteria containing hydrogenase reduce oxygen with hydrogen (the "Knallgas" reaction) faster than they reduce sulphate, because of the autoxidizable nature of cytochrome C_3 (Postgate, 1956, 1965).

For sulphate-reducing bacteria, anaerobiosis is usually achieved by growing cultures, (a) in vessels in which the air is replaced by an oxygen-free gas, (b) in evacuated vessels, (c) in vessels containing alkaline pyrogallol or other oxygen-absorbing chemicals, or (d) in airtight vessels completely filled with liquid or agar media. The choice of method depends often on the facilities available for gassing etc, as there have been few, if any, comparisons of the effectiveness of the different methods. Hydrogen, nitrogen or mixtures of these gases with 1–5% carbon dioxide (by volume) are often used to replace air. Hydrogen probably encourages growth of bacteria containing hydrogenase; strains that fix nitrogen will, of course, grow in a medium deficient in nitrogen compounds only if gaseous nitrogen is present. Carbon dioxide may encourage growth for the reasons given on p. 225 especially in media in which there is only a little organic carbon. Carbon dioxide also helps sometimes to buffer the pH, and to displace sulphide from solution; a gas phase consisting of 99% carbon dioxide and 1% hydrogen (v/v) has been used recently in studies of sulphate-reducing bacteria in anaerobic raw sewage sludge (Toerien, Thiel and Hattingh, 1968).

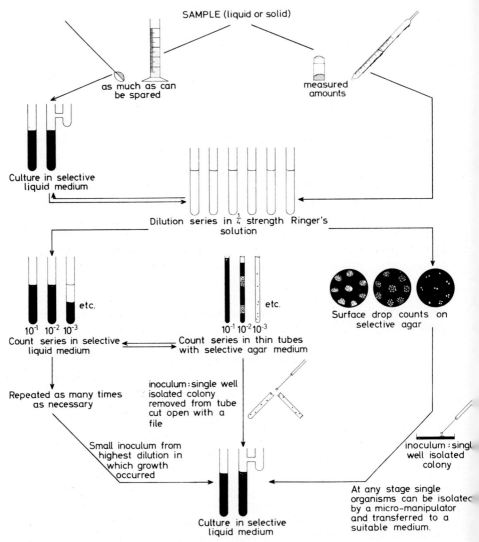

SAMPLE (liquid or solid)

as much as can be spared

measured amounts

Culture in selective liquid medium

Dilution series in $\frac{1}{4}$ strength Ringer's solution

etc.

10^{-1} 10^{-2} 10^{-3}
Count series in selective liquid medium

10^{-1} 10^{-2} 10^{-3}
Count series in thin tubes with selective agar medium

Surface drop counts on selective agar

Repeated as many times as necessary

inoculum: single well isolated colony removed from tube cut open with a file

Small inoculum from highest dilution in which growth occurred

inoculum: single well isolated colony

At any stage single organisms can be isolated by a micro-manipulator and transferred to a suitable medium.

Culture in selective liquid medium

FIG. 1. Some methods for isolating sulphate-reducing bacteria.

N.B. These procedures should be repeated until the culture appears to be pure microscopically; the culture should then be tested for contaminants (see p. 223). Mesophilic strains should be isolated at 30°, rumen strains at 37°, and thermophilic strains at 55°. Strains that do not form spores are killed by exposing cultures to 100° for 10 min.

Detection and Isolation

Some procedures for detecting and isolating sulphate-reducing bacteria are shown diagrammatically in Fig. 1. Whichever sequence of methods is chosen, the main requirement is patience. Growth on agar plates is better if the surface of the medium is inoculated with drops of a dilute suspension of cells, rather than by streaking.

Material suspected to contain sulphate-reducing bacteria should be transferred into a selective medium. I think that Baars' medium supplemented with yeast extract and sodium thioglycollate (see Table 1) is suitable for most purposes. The sediment it contains simulates conditions often found in the natural habitats of sulphate-reducing bacteria, with high ratios of solids to liquid; this initial sediment seems to provide a favourable micro-environment with a low E_h, as *D. vulgaris* starts to grow within and other strains grow round it as satellite-like colonies. Growth is accompanied by precipitation of ferrous sulphide, which causes the medium to blacken. If samples are from a salt-water environment, or halophilic organisms are known to be present, the medium should preferably contain 2·5% (w/v) NaCl. Although it was once thought that a requirement for salt, or ability to grow in a medium of a certain salinity, was characteristic of particular species, Miller, Hughes, Saunders and Campbell (1968) found that marine and fresh-water strains of *D. desulfuricans* and *D. vulgaris* rapidly adapted to growth either in the absence or presence of NaCl.

The quicker the samples are introduced to the appropriate medium the better, especially if there are only a few organisms present. Sulphate-reducing bacteria usually survive well if sample vessels are completely filled and tightly stoppered or closed, or if anaerobic conditions are maintained in some other way. If there is a delay in examining samples, they should be kept at about 4°, this is especially important if counts are to be done.

Culture tube

Particularly useful for growing pure or impure cultures of sulphate-reducing bacteria are the tubes illustrated in Fig. 2 (Pankhurst, 1966a, 1967b). Stock strains have been maintained in similar tubes for 6 years. The advantages are:

(1). Any individual culture can be examined without disturbing the others; this is impossible with cultures incubated in anaerobic jars. (2). The tube can be opened to remove samples and sealed again after the addition of fresh alkaline pyrogallol. In any case, the alkaline pyrogallol should be removed after a month or two, because its volume increases as it picks up

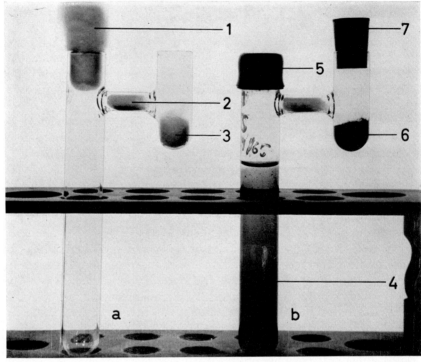

FIG. 2. A simple tube for isolating and cultivating sulphate-reducing bacteria.
Preparation. Tube **a** is ready for sterilization in an autoclave or oven. (1), non-absorbent cotton-wool plug; (2) loose plug of non-absorbent cotton wool, placed in position by a bent wire, and (3) wad of **absorbent** cotton wool.

In use. Tube **b**: (4), 10–20 ml medium containing a 3-day culture of *Desulfovibrio vulgaris*; (5), sterile bung or Suba-seal, replaces cotton-wool plug (1) immediately after inoculation of medium; (6), 0·5–1·0 ml 40 % (w/v) aqueous solution of pyrogallol and similar volume of saturated solution of sodium carbonate, placed on the cotton wool, and (7), rubber bung, unsterilized, as contents of side-arm are not sterile.

water vapour from the gas phase above the more dilute medium; there is then a danger that some of it may run over into the main tube. (3). If necessary, culture can be withdrawn by a syringe inserted through the Suba-seal. (4). Individual tubes and their atmospheres do not influence each other. (5). The space required for incubation is small; a single tube can be set up if necessary. (6). Access to a gas or a vacuum line is unnecessary. (7). The tubes are stable in ordinary test tube racks for $\frac{3}{4}$ in. diam tubes, and are no more fragile or less easy to clean than ordinary test tubes.

In order to follow oxygen uptake by alkaline pyrogallol in the side-arm, a tube was modified so that samples of its atmosphere could be withdrawn by syringe. The samples were analysed by gas chromatography. The oxygen concentration fell rapidly during the first five hours, and had this rate of fall continued there would have been none left after 7 or 8 h. However, there was still 1% by volume after 24 h. The absorption of the residual oxygen obviously takes longer, although air entering the sampling syringe or small leaks in the punctured Suba-seals may have accounted for some of this oxygen. Nevertheless, cultures rarely fail to grow within a day or two after inoculation, although it is probably advisable to use large inocula, 0·5–1·0 ml.

Pure cultures should be checked regularly for contamination by the rough agar dilution method (Postgate, 1953), or by incubating drops of culture anaerobically and aerobically on nutrient agar and in nutrient broth, and by examining hanging drop preparations or stained slides for typical morphology. In the rough agar dilution method, a sealed Pasteur pipette is dipped into the culture, and then dipped successively into seven thin test tubes containing a suitable molten agar at c. 40°. After the agar has set, the tubes are incubated at 30° and then examined for contaminant colonies, i.e. those colonies that are not black.

Enumeration

Viable sulphate-reducing bacteria are usually counted by a most probable number (MPN) technique in liquid media, or by colony counts in solid or semi-solid media. I have also found a surface drop method very useful, but have been unable to obtain comparable and reproducible results with the micro-techniques I have tried so far. Membrane filtration methods are also unsuccessful with sulphate-reducing bacteria (Postgate, 1969b).

Sulphate-reducing bacteria tend to form clumps containing ferrous sulphide, mucoid material and other micro organisms; even in pure cultures, old cells associate with precipitated ferrous sulphide. These clumps make microscopic and nephelometric determinations of total numbers of organisms meaningless or impossible, and also make estimations of viable organisms difficult. But organisms in samples containing micro-bially-produced sulphide (usually those samples which also contain numerous sulphate-reducing bacteria) are protected by the low E_h value; furthermore, the sulphide helps to establish a low initial E_h in the medium into which they are introduced. The absence of a reducing agent is often responsible for the phenomenon of "skipping"; sulphate-reducing bacteria grow in high dilutions of a sample, but not in low ones in which oxidizing chemicals are present.

The exact procedure for each method can be varied, and the choice of medium is often a matter of personal preference. I think that Baars' medium with yeast extract and sodium thioglycollate gives the best results. Regardless of method, the sample containing the organisms must be diluted, so that a wide range of population sizes is covered.

Dilutions

Samples are diluted serially; 1 ml or 1 g is introduced into 9 ml of sterile quarter strength Ringer's solution, and after careful mixing, 1 ml of this 10^{-1} dilution is transferred into a further 9 ml of diluent, and so on until an appropriate dilution (say 10^{-6}, 10^{-7} or 10^{-8}) has been reached. The Ringer's solution is sterilized by autoclaving just before use, and cooled rapidly under tap water to prevent re-solution of oxygen; 9 ml volumes are then pipetted into sterile tubes which are maintained at the optimum temperature for the organism concerned, to minimize death of organisms by thermal shock. Each dilution is prepared with a fresh pipette. Mixing of diluent and inoculum, and pipetting operations, are carried out with great care, to avoid unnecessary aeration of the liquid.

Most probable number method (liquid medium)

Allred, Mills and Fisher (1954) and Senez, Pichinoty and Geoffray (1956) described media for counting sulphate-reducing bacteria in secondary recovery systems and gasholder waters respectively. Grossman and Postgate (1953b) described an MPN procedure which was modified later (Drummond and Postgate, 1955) so that sulphate-reducing bacteria could be estimated in the presence of greater numbers of organisms that decompose cysteine.

Sets of five tubes, each containing 10 ml medium, are inoculated with 1g or 1 ml amounts of the original sample and with 1 ml volumes from appropriate dilutions. The original non-absorbent cotton-wool plug in each tube is trimmed and pushed down to a level just above the medium. A non-sterile absorbent cotton-wool plug, placed above but not touching the first plug, is soaked with 10 drops of a 40% (w/v) aqueous solution of alkaline pyrogallol and 10 drops of a saturated aqueous solution of sodium carbonate. The tube is then quickly closed with a rubber bung or Subaseal. The tubes are incubated in air, but oxygen is removed from the atmosphere inside the tubes by the alkaline pyrogallol, which if made up with sodium carbonate does not absorb carbon dioxide—for additional details, see this volume p. 65.

Tubes are examined from time to time for blackening. After 28 days or longer, the distribution of positive tubes (i.e. those that have blackened) at each dilution level is recorded. The most probable number of bacteria in the original sample is determined by reference to the statistical tables of Halvorsen and Ziegler (1933) or of Taylor (1962) (both given in Meynell and Meynell, 1965), or to McCrady's Probability Tables (Report, 1956; Postgate, 1969b). The exact incubation time is not critical; usually the results do not alter much after the first eight or nine days. An example follows:

Inoculum

	Undiluted sample	10^{-1} dilution	10^{-2} dilution
No. of positive tubes (out of five)	5	4	2

In this case, from Taylor's tables, there were 22 sulphate reducers/ml (or /g, if solid) of original sample.

Thin tube method (solid medium)

Several authors have described the use of narrow bore glass tubes for colony counts and for isolation of single colonies. Small inocula (usually 0·1 ml) from appropriate dilutions are introduced into sterile, thin-walled soda glass tubes (150 mm × 10 mm) with cotton-wool plugs. Duplicate tubes are inoculated at each dilution level. Each tube is almost filled with molten medium at about 40°, and the cotton-wool plug is discarded and replaced immediately by a sterile rubber bung. The tube is inverted twice to mix the inoculum and the medium before the latter sets.

The sealed tubes are incubated in air until well-defined black colonies are visible. This takes usually from 2–14 days. The tubes must be inspected regularly, as a disadvantage of counting black colonies within agar is that individual colonies are masked within a day or so of development by blackening of the entire medium. When no more colonies develop, the black ones are counted; care should be taken to avoid counting contaminant colonies, which will blacken if the tubes are left too long. The number of sulphate-reducing bacteria in the original sample is calculated as follows:

For example, with an inoculum of 0·1 ml the results might be:

Tubes inoculated with 10^{-1}, 10^{-2} and 10^{-3} dilutions: too many colonies to count.

Tubes inoculated with the 10^{-4} dilution: 1st tube, 16 colonies, 2nd tube, 22 colonies.

Tubes inoculated with the 10^{-5} dilution: 1st tube, 2 colonies, 2nd tube, 3 colonies.

The average number of colonies in tubes inoculated with the 10^{-4} dilution was 19. The volume of the inoculum was 0·1 ml; therefore there were $1·9 \times 10^6$ sulphate-reducing bacteria/ml or g of the original sample.

Surface drop method (solid medium)

The surface drop method is similar to the one described by Miles and Misra (1938). Drops (0·02 ml) of appropriate dilutions are placed on the surface of agar in Petri dishes, 8–10 drops/plate, one plate per dilution. Agar plates are prepared several days beforehand and refrigerated until needed; fresh plates are unsatisfactory, as drops take too long to dry. Stored plates can be incubated for a day at 30° before use, so that the drops are quickly absorbed and dry within 5–15 min. Inoculated plates (lids uppermost) are placed in sets of five or six in vacuum safety desiccators. Sealed desiccators are evacuated with a water pump to c. 30 mm Hg pressure; to remove any oxygen remaining after evacuation, each desiccator contains absorbent cotton wool soaked with 10 ml of an aqueous 40% (w/v) solution of pyrogallol and 10 ml of an aqueous saturated solution of sodium carbonate.

After incubation for 7–14 days, the desiccators are opened. The colonies on the plates are counted and the number of organisms/ml in the original suspension is calculated. Whenever possible c. 100 colonies are counted, to give an accuracy of $\pm 20\%$ (Badger and Pankhurst, 1960). If cultures are mixed, only colonies that are still surrounded by black zones when the plates have been exposed to air for an hour or two are counted. The results might be as follows:

Plate inoculated with the 10^{-2} dilution: colonies too crowded to count.
Plate inoculated with the 10^{-3} dilution: 10, 11, 15, 14, 15, 12, 13, 15 colonies/drop.
Plate inoculated with the 10^{-4} dilution: 1, 1, 0, 2, 0, 1, 1, 2 colonies/drop.

The total number of colonies counted on the 10^{-3} plate was 105, the average number/drop 13. Therefore there were $6·5 \times 10^5$ sulphate-reducing bacteria/ml or/g of the original sample.

Discussion

In simultaneous counts of D. vulgaris, the surface drop method gives results similar to those given by the MPN method, provided that the plates are incubated in genuinely anaerobic conditions (Pankhurst, 1967a). Some results of comparisons are shown in Fig. 3. The line drawn between the points represents the theoretical 1:1 relationship between the two methods. Although there is considerable scatter of points about this line, neither method gives consistently lower results than the other. The surface drop method seems as reliable as the MPN method, although the point on the graph where the surface drop figure is much lower illustrates a potential weakness in this method; this low result was caused, almost certainly, by the presence of oxygen in the desiccator in which the plates were incubated. The initial evacuation of the desiccator may have been inadequate or a leak

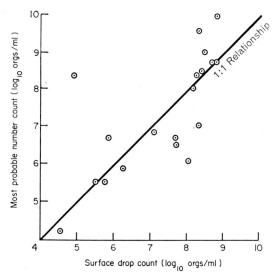

F IG. 3. The relationship between surface drop counts and most probable number counts for *D. vulgaris* strain Hildenborough (Baars' medium supplemented with yeast extract and sodium thioglycollate.)

may have developed later. When air is present *D. vulgaris* does not grow on the surface of agar plates but if conditions are first anaerobic and then become progressively aerobic, some growth can occur and lead to an inaccurate result. Under these circumstances, colony sizes in drops from the same dilution vary considerably and the difference between numbers of colonies developing in drops from the serial dilutions is greater than tenfold. In a satisfactory surface drop count, the numbers of colonies developing in drops from the same dilution vary only slightly, colony sizes are similar and there are approximately tenfold differences between the numbers of colonies which develop from the different dilutions.

In simultaneous MPN, surface drop and thin tube counts, the thin tube method usually gives slightly higher results than do the other two methods, showing that it is probably more reliable, as underestimation of numbers of sulphate-reducing bacteria are more likely than overestimations. On the other hand, it is unsuitable for strains which form a lot of gas. Theoretically, colony counts in which one colony is usually derived from one cell are more accurate than MPN counts in liquid media, as in the latter there is no way of distinguishing between growth that has originated from one cell or growth that is derived from more than one cell. The surface drop method has some advantages, as agar plates can be prepared in advance and refrigerated until required, whereas not only is fresh medium needed for thin tube and MPN counts, but preparation and inoculation take longer.

References

ABD-EL-MALEK, Y. & RIZK, S. G. (1958). Counting of sulphate-reducing bacteria in mixed bacterial populations. *Nature, Lond.,* **182,** 538.

ABD-EL-MALEK, Y. & RIZK, S. G. (1960). Culture of *Desulfovibrio desulfuricans. Nature, Lond.,* **185,** 635.

ADAMS, M. E. & POSTGATE, J. R. (1959). A new sulphate-reducing vibrio. *J. gen. Microbiol.,* **20,** 252.

ALLRED, R. C., MILLS, T. A. & FISHER, H. B. (1954). Bacteriological techniques applicable to the control of sulphate-reducing bacteria in flooding operations. *Producers Mon.Penn. Oil Prod. Ass.,* **19,** 31.

AMERICAN PETROLEUM INSTITUTE. (1965). Recommended practice for biological analysis of subsurface injection waters. 2nd ed. AP RP38. Div. of Production, Dallas, U.S.A.

BAARS, J. K. (1930). Over sulfaat reductive door bacteriën. Dissertation, University of Delft, Holland.

BADGER, E. H. M. & PANKHURST, E. S. (1960). Experiments on the accuracy of surface drop bacterial counts. *J. appl. Bact.,* **23,** 28.

BEIJERINCK, M. W. (1895). Über *Spirillum desulfuricans* als Ursache von Sulfat Reduktion. *Zentbl. Bakt. ParasitKde,* Abt. II, **1,** 104.

BOOTH, G. H. & WORMWELL, F. (1961). Corrosion of mild steel by sulphate-reducing bacteria. Effect of different strains of organisms. *1st Int. Congress on metallic corrosion.* London: Butterworth.

BUFTON, A. W. J. (1959). A note on the enumeration of thermophilic sulphate-reducing bacteria (*Clostridium nigrificans*). *J. appl. Bact.,* **22,** 278.

BUNKER, H. J. (1939). Factors influencing the growth of *Vibrio desulphuricans. Abstr. Comm. 3rd Int. Congr. Microbiol.,* 64.

BUTLIN, K. R., ADAMS, M. E. & THOMAS, M. (1949). The isolation and cultivation of sulphate-reducing bacteria. *J. gen. Microbiol.,* **3,** 46.

CAMPBELL, L. L. & POSTGATE, J. R. (1965). Classification of the spore-forming sulphate-reducing bacteria. *Bact. Rev.,* **29,** 359.

COLEMAN, G. S. (1960). A sulphate-reducing bacterium from the sheep rumen. *J. gen. Microbiol.,* **22,** 423.

DELDEN, A. VAN. (1903). Beitrag zur Kenntnis der Sulfatreduktion durch Bakterien. *Zentbl. Bakt. ParasitKde,* Abt. II, **11,** 81.

DRUMMOND, J. P. M. & POSTGATE, J. R. (1955). A note on the enumeration of sulphate-reducing bacteria in polluted water and on their inhibition by chromate. *J. appl. Bact.,* **18,** 307.

FISHER, H. F., KRASNA, A. I. & RITTENBERG, D. (1954). The interaction of hydrogenase with oxygen. *J. biol. Chem.*, **209**, 569.

GROSSMAN, J. P. & POSTGATE, J. R. (1953a). Cultivation of sulphate-reducing bacteria. *Nature, Lond.*, **171**, 600.

GROSSMAN, J. P. & POSTGATE, J. R. (1953b). The estimation of sulphate-reducing bacteria (*D. desulfuricans*). *Proc. Soc. appl. Bact.*, **16**, 1.

HALVORSEN, H. O. & ZIEGLER, N. R. (1933). Quantitative bacteriology. *J. Bact.*, **26**, 559.

HEWITT, L. F. (1950). *Oxidation-Reduction Potentials in Bacteriology and Biochemistry*, 6th ed. Edinburgh: E. & S. Livingstone Ltd.

HOBERMAN, H. D. & RITTENBERG, D. (1943). Biological catalysis of the exchange reaction between water and hydrogen. *J. biol. Chem.*, **147**, 211.

IVERSON, W. P. (1966). Growth of *Desulfovibrio* on the surface of agar media. *Appl. Microbiol.*, **14**, 529.

IVERSON, W. P. (1967). Disulfur monoxide: production by *Desulfovibrio*. *Science, N.Y.*, **156**, 1112.

MACPHERSON, R. & MILLER, J. D. A. (1963). Nutritional studies on *Desulfovibrio desulfuricans* using chemically defined media. *J. gen. Microbiol.*, **31**, 365.

MECHALAS, B. J. & RITTENBERG, S. C. (1960). Energy coupling in *Desulfovibrio desulfuricans*. *J. Bact.*, **80**, 501.

MEYNELL, G. G. & MEYNELL, E. (1965). *Theory and Practice in Experimental Bacteriology*. Cambridge: The University Press.

MILES, A. A. & MISRA, S. S. (1938). The estimation of the bactericidal power of the blood. *J. Hyg., Camb.*, **38**, 732.

MILLER, L. P. (1949). Rapid formation of high concentrations of hydrogen sulphide by sulphate-reducing bacteria. *Contr. Boyce Thompson Inst. Pl.Res.*, **15**, 437.

MILLER, J. D. A., HUGHES, J. E., SAUNDERS, G. F. & CAMPBELL, L. L. (1968). Physiological and biochemical characteristics of some strains of sulphate-reducing bacteria. *J. gen. Microbiol.*, **52**, 173.

PANKHURST, E. S. (1966a). Method and apparatus for growing anaerobic cultures. Brit. Patent 1, 041, 105.

PANKHURST, E. S. (1966b). The effect of gaseous environment on viable counts of *Desulfovibrio desulfuricans* and on sulphate reduction by washed cells. *Abstracts of Proceedings, IX international Congress for Microbiology*. Moscow, 1966, 146.

PANKHURST, E. S. (1967a). *The growth and occurrence of sulphate-reducing bacteria, with particular reference to their importance in the gas industry*. Ph.D. Thesis, University of London.

PANKHURST, E. S. (1967b). A simple culture tube for anaerobic bacteria. *Lab. Pract.* **16**, 58.

POSTGATE, J. R. (1951). On the nutrition of *Desulfovibrio desulfuricans*. *J. gen. Microbiol.* **5**, 714.

POSTGATE, J. R. (1953). On the nutrition of *Desulfovibrio desulfuricans*: a correction. *J. gen. Microbiol*, **9**, 440.

POSTGATE, J. R. (1956). Cytochrome C_3 and desulfoviridin; pigments of the anaerobe *Desulfovibrio desulfuricans*. *J. gen. Microbiol.*, **14**, 545.

POSTGATE, J. R. (1959). Sulphate reduction by bacteria. *A. Rev. Microbiol.*, **1**, 505.

POSTGATE, J. R. (1960). On the autotrophy of *Desulfovibrio desulfuricans*. *Z.allg. Microbiol.*, **1**, 53.

POSTGATE, J. R. (1965). Recent advances in the study of the sulphate-reducing bacteria. *Bact. Rev.*, **29**, 425.

POSTGATE, J. R. (1966). Media for sulphur bacteria. *Lab. Pract.*, **15**, 1239.

POSTGATE, J. R. (1969*a*). Media for sulphur bacteria: some amendments. *Lab. Pract.*, **18**, 286.

POSTGATE, J. R. (1969*b*). Viable counts and viability. In *Methods in Microbiology*. **1**, 611. Ed. Norris & Ribbons. London and New York: Academic Press.

POSTGATE, J. R. & CAMPBELL, L. L. (1966). Classification of *Desulfovibrio* species, the non-sporulating sulphate-reducing bacteria. *Bact. Rev.*, **30**, 732.

REPORT. (1956). The bacteriological examination of water supplies. *Rev. publ. Hlth. med. Subj. Lond.*, No. 71, 3rd ed. London, H.M.S.O.

ROGERS, T. H. (1940). The inhibition of sulphate-reducing bacteria by dyestuffs. *J. Soc. chem. Ind., Lond.*, **59**, 34.

SENEZ, S. C., PICHINOTY, F. & GEOFFRAY, C. (1956). Rôle des bactéries sulfate-réductrices dans la pollution des gazomètres, *C.r. 73ᵉ Congr. Ass. tech. Ind. Gaz,* 542.

SOROKIN, YU. I. (1961). Utilisation of the carbon in carbon dioxide by *Vibrio desulfuricans* and certain heterotrophic bacteria. *Dokl. Akad. Nauk. SSSR.*, **132**, 464. (Cited by Postgate, 1965).

SOROKIN, YU. I. (1966). Utilisation of carbon dioxide in biosynthesis by microorganisms. *Proc. Intern. Congr. Biochem., 5th, Moscow.*, **22**, 453.

SOROKIN, YU. I. (1966). Role of carbon dioxide and acetate in biosynthesis by sulphate-reducing bacteria. *Nature, Lond.*, **210**, 551.

STARKEY, R. L. (1938). A study of spore formation and other morphological characteristics of *Vibrio desulfuricans*. *Arch. Mikrobiol.*, **9**, 268.

STARKEY, R. L. & WIGHT, K. M. (1945). Anaerobic corrosion of iron in soil. *Amer. Gas Assoc. Proc. 27th Ann. Meeting.* New York: Amer. Gas Assoc. Inc.

TAYLOR, J. (1962). The estimation of numbers of bacteria by tenfold dilution series. *J. appl. Bact.*, **25**, 54.

TOERIEN, D. F., THIEL, P. G. & HATTINGH, M. M. (1968). Enumeration isolation and identification of sulphate-reducing bacteria of anaerobic digestion. *Water Res.*, **2**, 505.

ZELINSKY, N. D. (1893). On hydrogen sulphide fermentation in the Black Sea and the Odessa estuaries. *Proc. Russ. Phys. Chem. Soc.*, **25**, 298 (Cited by Baars, 1930).

ZOBELL, C. (1958). The ecology of sulphate-reducing bacteria. *Producers Mon. Penn. Oil Prod. Ass.*, **22**, 12.

Enrichment and Isolation of Photosynthetic Bacteria

R. Whittenbury

*Department of General Microbiology, University of Edinburgh,
(School of Agriculture), Edinburgh, Scotland*

Photosynthetic bacteria exist in anaerobic marine and fresh water environments. They use organic and inorganic electron donors. Enrichment and isolation of these unique organisms is, in many instances, a tedious process requiring patience, skill and ingenuity. Many people, Professor C. B. van Niel and Dr N. Pfennig in particular, have contributed to the development of the methods and media used in the isolation of photosynthetic bacteria described here.

Groups of Photosynthetic Bacteria

Three major groups of photosynthetic bacteria have been recognized and enrichment and isolation methods differ for each group.

Group 1

The purple, brown and brown-green non-sulphur-bacteria (*Athiorhodaceae*).

All use organic compounds as electron donors, are inhibited by H_2S and most are capable of aerobic heterotrophic growth in the dark. All strains probably use H_2 as an electron donor and fix CO_2 autotrophically.

Group 2

The purple, violet, pink, peach-coloured and red sulphur-bacteria (*Thiorhodaceae*).

All use H_2S (some $S_2O_3^=$) as an electron donor and fix CO_2 autotrophically. All are strict anaerobes and most are able to photoassimilate organic compounds.

Group 3

The green and brown-green sulphur-bacteria (*Chlorobacteriaceae*).

All use H_2S as an electron donor (some use $S_2O_3^=$) and fix CO_2 auto-trophically. All are strict anaerobes and predominantly photoautotrophs. Acetate improves the growth rate and yield of many but is not used in the absence of CO_2.

Enrichment

Enrichment systems are devised by juggling with a number of parameters, principally the type and concentration of substrate, pH value, temperature, light intensity and wave length, NaCl concentration, and choice of habitat. Latterly the use of different concentrations of substrate (H_2S in particular), infra-red filters and different pH values have enabled species to be isolated for the first time (Pfennig, 1967). A number of enrichment systems are described below including a non-specific one useful for teaching purposes.

Simple technique for obtaining photosynthetic bacteria from muds

The enrichment vessel, as described in Fig. 1, is placed 60 cm from a tungsten bulb (60 watt) and held at 30°. Oxygen in the system is removed by aerobic heterotrophs growing on the dissolved organic compounds

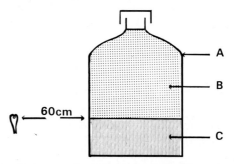

FIG. 1. Enrichment of photosynthetic bacteria. A, Bottle or glass vessel of size and dimension depending on availability. A round bottle, screw cap, of about 250 ml capacity is ideal for this purpose. B, Tap water containing (w/v); $MgSO_4.7H_2O$, 0·01%; KH_2PO_4, 0·1%; NH_4Cl, 0·05%; Fe EDTA, 0·0004%, pH 6·8–7·2. C, Mud from pond, ditch or other source, preferably black (indicating H_2S production) mixed with $CaSO_4$, 0·1% and cellulose powder 0·5%. In many instances sufficient H_2S is generated in the absence of additional $CaSO_4$. Cellulose can be omitted or replaced by lactate or other organic acid. The mud mixture occupies about 25% of the vessel volume, the tap water-salts mixture about 70% of the volume. The whole is gassed with N_2 for about 5 min to lower the O_2 content.

diffusing from the mud or on the lactate (if added). Cellulose when present is utilized by cellulolytic clostridia, reduced organic compounds, H_2 and CO_2 being formed. At the onset of gas production the screw-cap is loosened daily to avoid an excessive build-up of pressure. Reduced organic compounds (and any added lactate) and hydrogen are now used as electron donors by sulphate-reducing bacteria to form H_2S. The generation of H_2S is first indicated by a blackening of the mud (the formation of FeS). As hydrogen sulphide and CO_2 continues to be generated, the sulphur purple and green bacteria make their appearance. The first signs of their presence are usually seen at the mud-glass interface and in a line down the back of the bottle where light is focused by the bottle contents. Colonies then begin to appear on the glass surface above the mud and the dense growth of organisms often deeply colour the medium. Occasionally the enrichment may fail at this stage; the generation of H_2S may exceed utilization by the photosynthetic bacteria and the whole culture blackens masking the presence of photosynthetic bacteria. In most instances, however, photosynthetic bacteria utilize the H_2S rapidly. Both green and purple varieties may appear depending upon the pH and H_2S concentration. In addition non-sulphur purple bacteria are enriched; they use H_2 and reduced organic compounds and can be recognized by microscopial examination.

As a source of isolates this enrichment system is not ideal because the conditions finally produced by the microbial associations are not known. However, it is often possible to isolate the more commonly occurring photosynthetic bacteria from these enrichments, such as rhodopseudo-monads, the small *Chromatium* spp and some *Chlorobium* spp.

Specific enrichments and isolation methods

Enrichment and isolation of non-sulphur-utilizing bacteria

Rationalizing enrichment media and conditions led Van Niel (1944) to devise selective systems for the enrichment and isolation of *Rhodopseudomonas* and *Rhodospirillum* spp. An extension of Van Niel's methods (see Tables 1, 2 and 5 for medium and selective conditions) has led recently to the isolation of new species with unusual properties, such as the green *Rhodopseudomonas viridis*, Drews and Giesbrecht (1966) and the red *Rhodopseudomonas acidophila*, Pfennig (1969a), which forms sessile buds. Isolation of these organisms is usually by the agar shake method. A portion of the enrichment culture is serially diluted in a liquefied agar medium which, after gelling, is overlaid to *c.* 2 cm with sterile medium. The tubes are placed in front of a light. Many of these organisms are initially very sensitive to oxygen (but after a number of subcultures many become

s

TABLE 1. Major components of medium for non-sulphur-photosynthetic bacteria

Constituent	Constituents /litre distilled water	Comments
NH$_4$Cl	0·5g	
MgSO$_4$.7H$_2$O	0·4g	Prior to autoclaving, add to
NaCl	0·4g	tubes or bottles (melt agar
CaCl$_2$.2H$_2$O	0·05g	before doing so). Autoclave (15
Electron donor		min at 121°), cool, add other
(i.e. succinate)	2·0g–0·2g	heat-labile materials if neces-
Agar (if required)	15·0g	sary.
Yeast extract	1·0g	
(if required)		
KH$_2$PO$_4$	1·0g	
Trace element mixture	10ml	Filter-sterilize and add aseptic-
(See Table 5)		ally to autoclaved medium.
Growth factors (if required)		
(Table 3)		

Adjust pH as required, before autoclaving at 121°/15 min.

TABLE 2. Enrichment of non-sulphur-utilizing bacteria

Species	Enrichment conditions	Preferred method of isolation
A. Budding species		
Rhodopseudomonas palustris	Basal medium + yeast extract + 0·1–2% (w/v) ethanol, methanol, glutarate or benzoate; pH 6·8.	Agar shakes or surface streaking of agar plates.
Rhodopseudomonas viridis (Drews and Giesbrecht, 1966)	Basal medium + yeast extract + 0·2%, (w/v) succinate; infrared filter; pH 6·8.	Agar shakes.
Rhodopseudomonas acidophila (Pfennig, 1969a)	Basal medium + 0·2% (w/v) succinate, pH 5·0–5·5.	Agar shakes.
Rhodomicrobium vannielii	Basal medium, no yeast extract or growth factors. 0·03% (w/v) Na$_2$S.9H$_2$O; chloride salts replaced with sulphate salts; 0·2% (w/v) acetate, pH 6·8.	Agar shakes.
B. Non-budding species		
Rhodopseudomonas capsulatus	Basal medium + yeast extract + 0·2% (w/v) propionate, lactate or succinate or H$_2$ + CO$_2$ for photoautotrophic growth.	Agar shakes.

TABLE 2 *continued*

Species	Enrichment conditions	Preferred method of isolation
Rhodopseudomonas spheroides	**1st stage** Basal medium + yeast extract + 0·1–0·2% (w/v) alcohols or fatty acids (acetate, propionate), pH 6·5. **2nd stage** Transfer to basal medium + yeast extract + 0·2% (w/v) tartrate, pH 6·0.	Agar shakes.
Rhodopseudomonas gelatinosa	**1st stage** Basal medium + yeast extract + 0·2% (w/v) citrate, pH 6·5. **2nd stage** Transfer to basal medium + yeast extract + 0·2% (w/v) citrate.	Agar shakes.
Rhodospirillium fulvum	Basal medium + yeast extract + 0·04% (w/v) pelargonate or caprylate (p-aminobenzoic acid can replace yeast extract).	Agar shakes. Ascorbate (0·05% w/v) included as these organisms are strict anaerobes.
Rhodospirillium rubrum	**1st stage** Basal agar medium + yeast extract + 0·1% (v/v) methanol, ethanol or L-alanine. **2nd stage** Transfer to basal medium + yeast extract + 0·1% (v/v) ethanol, pH 6·8.	Agar shakes. Ascorbate (0·05% w/v) included.
Rhodospirillium tenue (Pfennig, 1966b)	Basal medium, no yeast extract or growth factors, + 0·1% (w/v) pelargonate, acetate or propionate, pH 6·8.	Agar shakes.

oxygen-tolerant) and ascorbate or other reducing agent is an essential ingredient in the medium. Three or four transfers through agar shakes usually ensures a pure culture which can be kept in agar stabs in sealed vials for months between transfers. Transfers are made by blowing the agar out of a tube into a Petri dish by nitrogen gas through a Pasteur pipette inserted down the side of the culture. Colonies are picked and transferred. Occasionally it is difficult to separate two or sometimes three species in an enrichment and a second stage enrichment may be necessary to promote the dominance of one type. Also, streaking on the surface of agar plates and exposing to light will separate the more aero-tolerant varieties, such as *Rhodopseudomonas palustris*. Examples of all these methods are given in Table 2.

Sulphur-utilizing bacteria

Hydrogen sulphide concentrations, temperature, light intensity and use of infra-red filters are the main variables used in the enrichment of sulphur-utilizing bacteria and, in the past few years, have resulted in the enrichment of some unusual organisms. As a group they are more difficult to handle than non-sulphur-utilizing bacteria; they are strict anaerobes and require media which are laborious to prepare (Table 3). In many instances, the organisms will not form colonies in agar thus presenting difficulties in culture and isolation. Consequently few people seem to have mastered the technique of isolating pure cultures of the larger organisms. Table 4 provides examples of selective systems successfully employed in enrichments. In all cases, round bottles filled with liquid medium almost to the top (a small gas space being left to allow for liquid expansion) are ideal.

The green species and small-celled *Chromatium* spp are usually isolated by serially diluting the enrichments in agar deeps, preferably in screw-cap tubes to lessen the risk of growth inhibition by oxygen. Exposure to tungsten light for several days at 30° results in the formation of brown – green colonies in the medium. Removal of the agar shake into a Petri dish (by nitrogen passed through a Pasteur pipette down the side of the agar shake), and the picking of several colonies into fresh agar-shakes diluting as before, is normally sufficient to ensure a pure culture. Picking of colonies from the second shakes into agar deeps or liquid medium in sealed bottles is the final process. The usual procedures are carried out to be quite certain that only photosynthetic microorganisms exist in the culture; the most frequent contaminant is usually a sulphate-reducing bacterium, commonly a *Desulfovibrio* spp, which lives on the organic products released by auto-lysing photosynthetic bacteria and reduces the sulphate formed by the photosynthetic bacteria.

As the larger organisms do not grow in agar shakes, other subterfuges are employed. The simplest is serial dilution in liquid medium, in the hope that the photosynthetic bacterium is the numerically dominant organism. Limitations of this technique are the difficulty of ensuring that only one type of photosynthetic bacterium is present and the difficulty of eliminating *Desulfovibrio* spp. If more than one type of photosynthetic bacterium is present, modifications of the growth medium based on enrichment

TABLE 3. Medium for sulphur-photosynthetic bacteria, constituents/litre distilled water

Constituent		Comments
$MgSO_4.7H_2O$	1·0g	
$CaSO_4.2H_2O$	0·5g	
$CaCO_3$	0·05g	Dissolve in 700ml distilled water,
$FeSO_4.7H_2O$	10·0mg	autoclave 15 min at 121° and cool
$(NH_4)_2SO_4$ or Ammonium acetate	1·0g	before adding other constituents.
KH_2PO_4	1·0g	Dissolve in 100ml distilled water, autoclave 15 min at 121° and add to cooled basal medium.
Trace element solution, (see Table 5)	10ml	Autoclave, cool and add to basal medium.
Growth factor solution, (see Table 5)	10ml	Filter-sterilize before adding to cooled basal medium.
Vitamin B_{12}	20mg	Dissolve in 10ml distilled water and filter-sterilize before adding to cooled basal medium.
Na_2S	0·5–0·05g	Amount dependent on organisms (Table 4). Generally added after partially neutralizing with dilute HCl (all aseptically), before adding aseptically to cooled bulk medium. Magnetic stirring and checking of pH between the additions usually used.
pH Value		Adjust as required (see Table 4); can be lowered by bubbling CO_2 through bulk medium immediately before dispensing in sterilized bottles (all aseptic procedures).
Dispense medium		In sterile bottles, leaving 2–3ml gas space to allow for expansion. A slight precipitate is usually formed. Allow to settle in cold room for 2–3 days before use.

techniques may encourage a change in the relative numbers of the types and so allow isolation of both by serial dilution.

Another method, used successfully for isolating large motile organisms but requiring skill and patience, is the capillary method derived from the method of Giesberger (1947). A Pasteur pipette is drawn and an area in the middle of the capillary is flattened so that its contents can be viewed microscopically. The pipette is sterilized, sterile medium drawn in and one end flame-sealed. A drop or two of the enrichment is introduced at the open end which is then sealed. The sealed pipette is mounted on the stage of a phase-contrast microscope. A watch is now kept on the flattened area of the tube and when a large organism has passed through the viewing zone the pipette is broken at this point, sealed and incubated in the light. A proportion of such cultures grow, and a few usually prove to be free of non-photosynthetic bacteria and contain just the one morphological type of photosynthetic bacterium.

Non-motile bacteria are normally only successfully isolated by serial dilution in liquid medium. However, one or two manipulations can be tried to enrich a particular organism. Gas-vacuole containing photosynthetic bacteria will accumulate under the bottle stoppers at low temperatures, 4–10°; opening the bottle and allowing air to enter promotes a negative response to oxygen by many of the larger motile organisms and they move rapidly to the bottom of the bottle, thereby increasing the relative numbers of floating organisms. Transfers of the latter organisms followed by serial dilutions may result in their successful isolation.

Millipore filters have been used successfully in this laboratory to increase the dominance of large organisms. Using a series of filters with decreasing pore-size, it is possible to retain the largest organisms on the filters and then to float them off before preparing serial dilutions. Using the one sample it is sometimes possible to enrich 2 or 3 large types and, from the

TABLE 4. Enrichment of sulphur-utilizing photosynthetic bacteria (all in same basal medium—Table 3, sulphide concentration, light intensity and incubation temperature are the major selective factors)

Selective conditions	Examples of organisms enriched
High sulphide concentration (e.g. 0·5% w/v); high light intensity;* pH 6·7–7·0, 25–30°.	Predominantly the green bacteria, e.g. *Chlorobium* spp.
High sulphide concentration, high light intensity* and infra-red filter, pH 7·0–7·5, 25–30°.	Purple bacteria, e.g. *Chromatium, Thiospirillium, Thiocystis, Thiocapsa* and *Thiococcus* spp.
Low sulphide concentration (e.g. 0·05%), low light intensity,† pH 6·7–7·5, low temperature (10–20°).	*Rhodothece, Amoebobacter, Thiodictyon, Lamprocystis, Pelodictyon* and *Chlathrochloris* spp.

Isolation (see text for details). Same medium as enrichment medium is used. Gas-vacuole forming varieties (*Lamprocytis, Rhodothece, Amoebobacter* and *Thiodictyon* spp.) float to the top of the cultures at low temperatures (4–10°) and can be isolated by serial dilution. Strains which grow in agar are isolated by agar shake method. Other strains isolated by serial dilution, by Millipore filter or, if motile, by capillary method.

　* 50–200 foot candles.

　† 5–20 foot candles.

final filtrate, small organisms. In all operations, however, anaerobic conditions have to be maintained so far as is possible by gassing (N_2, H_2 or other inert O_2-free gas) and by including a reducing agent in the suspension.

Once the organisms are in pure culture, they are usually maintained in the same medium as that used for enrichments. A solution of H_2S is added at intervals until the culture has reached such a density that a fresh transfer has to be made. The time between transfers can be extended by use of low light levels and low temperature.

The media (Tables 1, 3 and 5) are based on those devised and used by Dr N. Pfennig (private communication) and Pfennig and Lippert (1966).

TABLE 5. Trace element solution and growth factor solution. (Dissolve in litre distilled water. Adjust pH if required)

Trace element solution	
$FeSO_4.7H_2O$	200mg
$ZnSO_4.7H_2O$	10mg
$MnCl_2,4H_2O$	3mg
H_3BO_3	30mg
$CoCl_2.6H_2O$	20mg
$CuCl_2.2H_2O$	1mg
$NiCl_2.6H_2O$	2mg
$Na_2MoO_4.2H_2O$	500mg

Growth factor solution	
Biotin	0·1mg
Ca-pantothenate	0·1mg
p–aminobenzoic acid	1·0mg
Nicotinic acid	2·0mg
Thiamine hydrochloride	2·0mg

Dissolve in 100ml distilled water and filter-sterilize.

References

DREWS, G. & GIESBRECHT, P. (1966). *Rhodopseudomonas viridis* nov. spec., ein neu isoliertes, obligat phototrophes Bacterium. *Arch. Mikrobiol.*, **53**, 255.

GIESBERGER, G. (1947). Some observations on the culture, physiology and morphology of some brown-red *Rhodospirillium* speces. *Antonie van Leeuwenhoek, J. Microbiol-serol.*, **13**, 135.

PFENNIG, N. (1967). Photosynthetic Bacteria. *Ann. Rev. Microbiol.*, **21**, 285.

PFENNIG, N. (1969a). *Rhodopseudomonas acidophila* sp. n., a new species of the budding purple non-sulfur bacteria. *J. Bact.*, **99**, 597.

PFENNIG, N. (1969b). *Rhodospirillium tenue* sp. n., a new species of the purple non-sulfur bacteria. *J. Bact.*, **99**, 619.

PFENNIG, N. & LIPPERT, K. D. (1966). Uber das vitamin B_{12}—Bedürfnis phototropher Schwefelbacterien. *Arch. Mikrobiol.*, **55**, 245.

VAN NIEL, C. B. (1944). The culture, general physiology morphology and classification of the non-sulfur purple and brown bacteria. *Bact. Rev.*, **8**, 1.

Author Index

T

Subject Index